Erwin Dee Kord (Ed.)

Outline of Palau

Erwin Dee Kord (Ed.)

Outline of Palau

United Nations Trusteeship Council, Geography of Palau, Pacific Ocean

Solv

Imprint

Publisher:
Solv is a trademark of
International Book Market Service Ltd., 17 Rue Meldrum, Beau Bassin, 1713-01 Mauritius
Email: info@bookmarketservice.com
Website: www.bookmarketservice.com

Published in 2011

Printed in: U.S.A., U.K., Germany. This book was not produced in Mauritius.

ISBN: 978-613-8-95493-4

Contents

Outline_of_Palau 1
United_Nations_Trusteeship_Council 10
Philippines 13
Palau 41
Geography_of_Palau 55
Micronesia 57
Island_country 59
Mount_Ngerchelchuus 62
Babeldaob 63
Environment_of_Palau 64
Wildlife_of_Palau 65
States_of_Palau 65
Demographics_of_Palau 67
Melekeok 69
Politics_of_Palau 71
Oceania 72
President_of_Palau 77
House_of_Delegates 78
Foreign_relations_of_Palau 79
History_of_Palau 80
Economy_of_Palau 84

References

Article Sources and Contributors 88
Image Sources, Licenses and Contributors 91

Outline_of_Palau

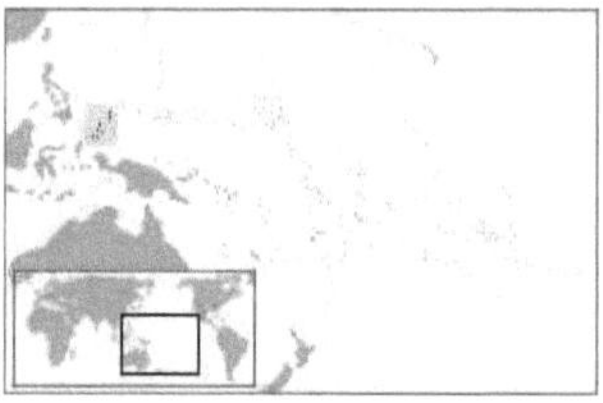

The location of Palau

The Flag of Palau

The Seal of Palau

The following outline is provided as an overview of and topical guide to Palau:

General reference

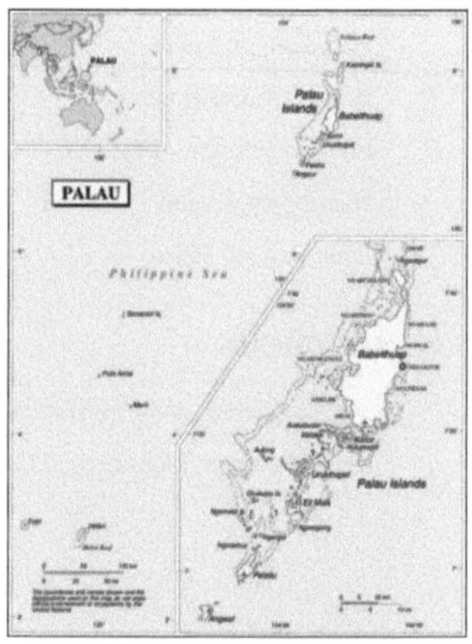

An enlargeable map of the Republic of Palau

- Pronunciation:
- Common English country name: Palau
- Official English country name: The Republic of Palau
- Common endonym(s): List of countries and capitals in native languages
- Official endonym(s): List of official endonyms of present-day nations and states
- Adjectival(s): Palauan
- Demonym(s):
- Etymology: Name of Palau
- ISO country codes: PW, PLW, 585
- ISO region codes: See ISO 3166-2:PW
- Internet country code top-level domain: .pw

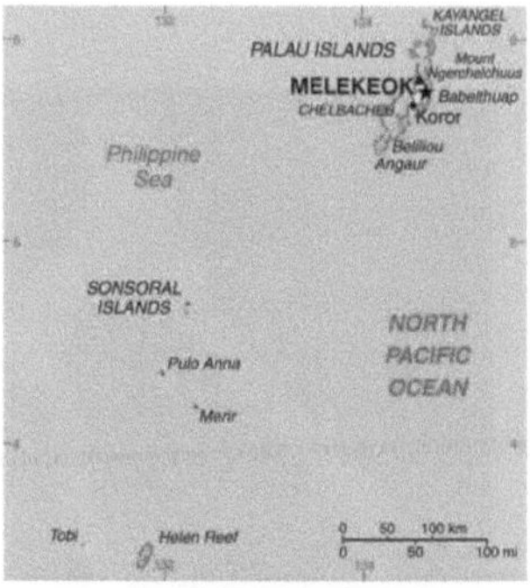

An enlargeable basic map of Palau

Geography of Palau

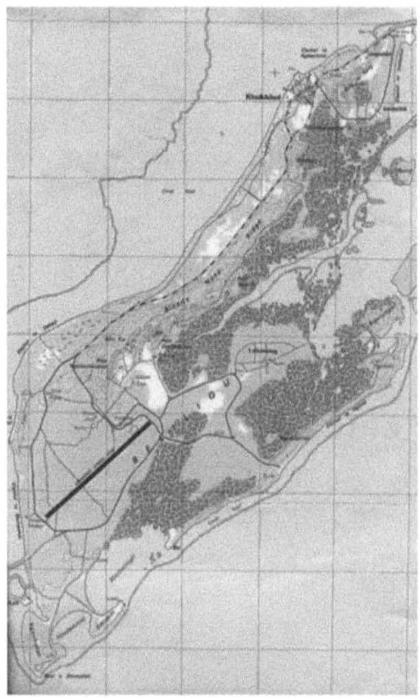
An enlargeable topographic map of the island of Peleliu

- Palau is: an island country
- Location:
  - Northern Hemisphere and Eastern Hemisphere
  - Pacific Ocean
    - North Pacific Ocean
      - Oceania
        - Micronesia
  - Time zone: UTC+09
  - Extreme points of Palau
    - High: Mount Ngerchelchuus on Babeldaob 242 m (794 ft)
    - Low: North Pacific Ocean 0 m
  - Land boundaries: *none*
  - Coastline: North Pacific Ocean 1,519 km
- Population of Palau: 20,000 - 211st most populous country
- Area of Palau: 459 km^2
- Atlas of Palau

Environment of Palau

- Climate of Palau
- Environmental issues in Palau
- Ecoregions in Palau
- Renewable energy in Palau
- Geology of Palau
- Protected areas of Palau
  - Biosphere reserves in Palau
  - National parks of Palau
- Wildlife of Palau
  - Flora of Palau
  - Fauna of Palau
    - Birds of Palau
    - Mammals of Palau

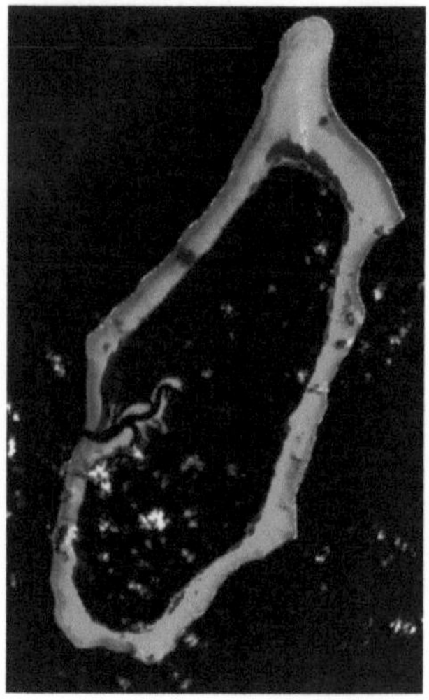

An enlargeable satellite image of Helen Reef in Palau

Natural geographic features of Palau

- Fjords of Palau
- Glaciers of Palau
- Islands of Palau
- Lakes of Palau

- Mountains of Palau
  - Volcanoes in Palau
- Rivers of Palau
  - Waterfalls of Palau
- Valleys of Palau
- World Heritage Sites in Palau: None

Regions of Palau

Administrative divisions of Palau

- States of Palau
  - Municipalities of Palau

States of Palau

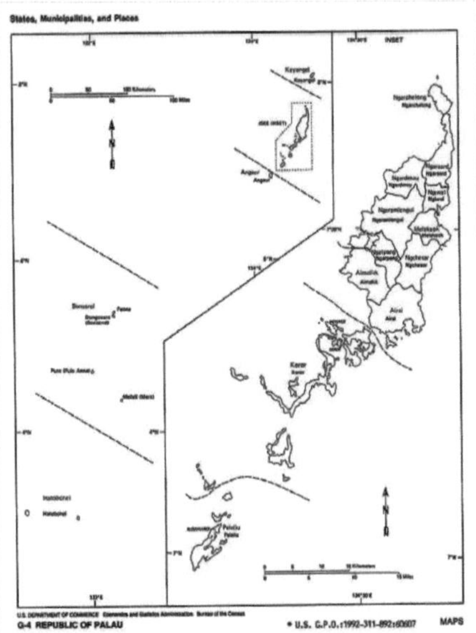

16 states of Palau

State	Capital	Area (km^2)	Population (Census 2000)
North of Babeldaob			
Kayangel	Kayangel	1	138
Babeldaob			
Aimeliik	Mongami	52	272
Airai	Ngetkib	44	2 104
Melekeok	Melekeok	28	239
Ngaraard	Ulimang	36	638
Ngarchelong	Mengellang	10	286
Ngardmau	Urdmang	47	221
Ngaremlengui	Imeong	65	367
Ngatpang	Ngereklmadel	47	280
Ngchesar	Ngersuul	41	267
Ngiwal	Ngerkeai	26	193

Southwest of Babeldaob			
Angaur	Ngeremasch	8	188
Koror	Koror	18	13 303
Peleliu	Kloulklubed	13	571
Rock Islands, (also called *Chelbacheb*)	-	47	-
Southwest Islands			
Hatohobei	Hatohobei	1	23
Sonsorol	Dongosaru	3	39
Palau	Melekeok[1]	444	19 129
[1] The national capital of Palau was moved from Koror to Melekeok on October 7, 2006.			

Municipalities of Palau

- Capital of Palau: Melekeok
- Cities of Palau

Government and politics of Palau

Main article: Government of Palau and Politics of Palau

- Form of government:
- Capital of Palau: Melekeok
- Elections in Palau
  - (specific elections)
- Political parties in Palau
- Political scandals of Palau
- Taxation in Palau

Branches of the government of Palau

- Executive Branch, consisting of one Head of Government
- Legislative Branch
  - Upper House:Senate
  - Lower House:House of Delegates
- Judiciary Branch

Executive branch of the government of Palau

- Head of state: President of Palau

Legislative branch of the government of Palau

- Parliament of Palau - Olbiil Era Kelulau (bicameral)
  - Upper house: Senate of Palau
  - Lower house: House of Delegates

Judicial branch of the government of Palau

- Supreme Court of Palau
- National Court
- Court of Common Pleas
- Land Court

Foreign relations of Palau

- Diplomatic missions in Palau
- Diplomatic missions of Palau
- United States-Palau relations

International organization membership

The Republic of Palau is a member of:[1]

- African, Caribbean, and Pacific Group of States (ACP)
- Asian Development Bank (ADB)
- Food and Agriculture Organization (FAO)
- International Atomic Energy Agency (IAEA)
- International Bank for Reconstruction and Development (IBRD)
- International Civil Aviation Organization (ICAO)
- International Development Association (IDA)
- International Federation of Red Cross and Red Crescent Societies (IFRCS)
- International Finance Corporation (IFC)
- International Monetary Fund (IMF)
- International Olympic Committee (IOC)
- International Red Cross and Red Crescent Movement (ICRM)
- Inter-Parliamentary Union (IPU)
- Multilateral Investment Guarantee Agency (MIGA)
- Organization for the Prohibition of Chemical Weapons (OPCW)
- Pacific Islands Forum (PIF)
- Secretariat of the Pacific Community (SPC)
- South Pacific Regional Trade and Economic Cooperation Agreement (Sparteca)
- United Nations (UN)
- United Nations Conference on Trade and Development (UNCTAD)
- United Nations Educational, Scientific, and Cultural Organization (UNESCO)
- World Health Organization (WHO)

Law and order in Palau

- Capital punishment in Palau
- Constitution of Palau
- Crime in Palau
- Human rights in Palau
  - LGBT rights in Palau
  - Freedom of religion in Palau
- Law enforcement in Palau

Military of Palau

- Command
  - Commander-in-chief:
    - Ministry of Defence of Palau
- Forces
  - Army of Palau
  - Navy of Palau
  - Air Force of Palau
  - Special forces of Palau
- Military history of Palau
- Military ranks of Palau

History of Palau

Main article: History of Palau, Timeline of the history of Palau, and Current events of Palau

- Economic history of Palau
- Military history of Palau

Culture of Palau

- Architecture of Palau
- Cuisine of Palau
- Ethnic minorities in Palau
- Festivals in Palau
- Humor in Palau
- Languages of Palau
- Media in Palau
- National symbols of Palau
  - Coat of arms of Palau
  - Flag of Palau
  - National anthem of Palau
- People of Palau
- Prostitution in Palau
- Public holidays in Palau
- Records of Palau
- Religion in Palau
  - Buddhism in Palau
  - Christianity in Palau
  - Hinduism in Palau
  - Islam in Palau
  - Judaism in Palau
  - Sikhism in Palau
- World Heritage Sites in Palau: None

Art in Palau

- Art in Palau
- Cinema of Palau
- Literature of Palau
- Music of Palau
- Television in Palau
- Theatre in Palau

Sports in Palau

- Football in Palau
- Palau at the Olympics

Economy and infrastructure of Palau

- Economic rank, by nominal GDP (2007): 186th (one hundred and eighty sixth)
- Agriculture in Palau
- Banking in Palau
  - National Bank of Palau
- Communications in Palau
  - Internet in Palau
- Companies of Palau
- Currency of Palau: Dollar
  - ISO 4217: USD
- Economic history of Palau
- Energy in Palau
  - Energy policy of Palau
  - Oil industry in Palau
- Mining in Palau
- Tourism in Palau
- Transport in Palau
- Palau Stock Exchange

Infrastructure of Palau

- Health care in Palau
- Transportation in Palau
  - Airports in Palau
  - Rail transport in Palau
  - Roads in Palau
- Water supply and sanitation in Palau

See also

- Index of Palau-related articles
- List of international rankings
- List of Palau-related topics
- Member state of the United Nations
- Outline of geography
- Outline of Oceania
- Palauan language

References

[1] "Palau" (https://www.cia.gov/library/publications/the-world-factbook/geos/ps.html). *The World Factbook*. United States Central Intelligence Agency. July 2, 2009. . Retrieved July 23, 2009.

External links

Wikimedia Atlas of Palau

- Official Site of the Republic of Palau (http://www.palaugov.net)
- Palau National Congress - Olbiil Era Kelulau, Senate (http://senate.palauoek.net)
- Palau (http://www.dmoz.org/Regional/Oceania/Palau/) at the Open Directory Project
- Outline of Palau travel guide from Wikitravel

United_Nations_Trusteeship_Council

United Nations Trusteeship Council

The chamber of the UN Trusteeship Council, UN headquarters, New York

Org type	Principal Organ
Head	*President* Michel Duclos France *Vice-President* Adam Thomson United Kingdom
Status	Inactive (As of 1994)
Established	1945
Website	[1]

The **United Nations Trusteeship Council**, one of the principal organs of the United Nations, was established to help ensure that trust territories were administered in the best interests of their inhabitants and of international peace and security. The trust territories—most of them former mandates of the League of Nations or territories taken from nations defeated at the end of World War II—have all now attained self-government or independence, either as separate nations or by joining neighbouring independent countries. The last was Palau, formerly part of the Trust Territory of the Pacific Islands, which became a member state of the United Nations in December 1994.

The world in 1945, UN Trusteeship territories are colored green

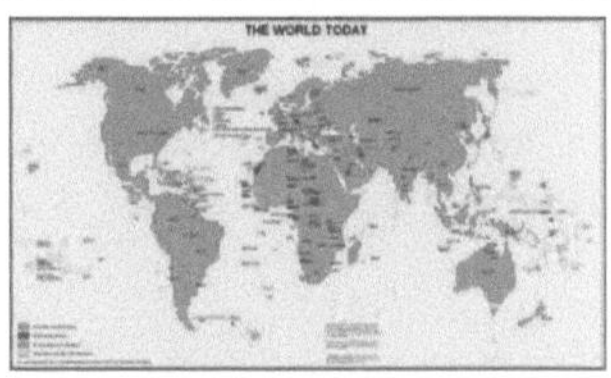

The world in 2000, with no trusteeship territories left

History

Arrival of UN Visiting Mission in Majuro, Trust Territory of the Pacific Islands (1978). The sign reads "Please release us from the bondage of your trusteeship agreement."

The Trusteeship Council was formed in 1945 to oversee the decolonization of those dependent territories that were to be placed under the international trusteeship system created by the United Nations Charter as a successor to the League of Nations mandate system. Ultimately, eleven territories were placed under trusteeship: seven in Africa and four in Oceania. Ten of the trust territories had previously been League of Nations mandates; the eleventh was Italian Somaliland.

Under the Charter, the Trusteeship Council was to consist of an equal number of United Nations Member States administering trust territories and non-administering states. Thus, the Council was to consist of (1) all U.N. members administering trust territories, (2) the five permanent members of the Security Council, and (3) as many other non-administering members as needed to equalize the number of administering and non-administering members, elected by the United Nations General Assembly for renewable three-year terms. Over time, as trust territories attained independence, the size and workload of the Trusteeship Council was reduced and ultimately came to include only the five permanent Security Council members (China, France, the Soviet Union/Russian Federation, the United Kingdom, and the United States).

With the independence of Palau, formerly part of the Trust Territory of the Pacific Islands, in 1994, there presently are no trust territories, leaving the Trusteeship Council without responsibilities. (Since the Northern Mariana Islands was a part of the Trust Territory of the Pacific Islands and became a commonwealth of the USA in 1986, it is technically the only area to have not joined as a part of another state or gained full independence as a sovereign nation.)

The Trusteeship Council was not assigned responsibility for colonial territories outside the trusteeship system, although the Charter did establish the principle that member states were to administer such territories in conformity with the best interests of their inhabitants.

Present status

Its mission fulfilled, the Trusteeship Council suspended its operation on 1 November 1994, and although under the United Nations Charter it continues to exist on paper, its future role and even existence remains uncertain. The Trusteeship Council is currently (as of 2005) headed by Michel Duclos, with Adam Thomson as vice-president,[1] although the sole current duty of these officers is to meet with the heads of other UN agencies on occasion. Initially they met annually, but according to a UN press release from their session in 2004:

> The Council amended its rules of procedure to drop the obligation to meet annually and agreed to meet as the occasion required. It now meets by its own decision, the decision of its President, at a request from a majority of its members, or at a request from the General Assembly or Security Council.

Future prospects

The formal elimination of the Trusteeship Council would require the revision of the UN Charter, which is why it has not been pursued. Other functions for the Trusteeship Council have been considered.

The Commission on Global Governance's 1994 report recommends an expansion of the trusteeship council. Their theory is that an international regulatory body is needed to protect environmental integrity and the global commons on the two-thirds of the world's surface that is outside national jurisdictions.[2]

In March 2005, however, UN Secretary-General Kofi Annan proposed a sweeping reform of the United Nations, including an expansion of the Security Council. As this restructuring would involve significant changes to the UN charter, Annan proposed the complete elimination of the Trusteeship Council as part of these reforms.[3]

References

[1] Trusteeship Council elects President, Vice-President; Adopts agenda for 64th session (http://www.un.org/News/Press/docs/2004/tr2426.doc.htm), United Nations Press Release TR/2426, 20 October 2004

[2] Shaw, John: UN Adviser Says World Must Focus On Sustainable Development (http://www.washdiplomat.com/01-11/a1_11_01.html), The Washington Diplomat

[3] Secretary-General's reflorm recommendations 'did not go far enough', general assembly told, as debate begins on 'In larger freedom' (http://www.un.org/News/Press/docs/2005/ga10337.doc.htm) UN Press Release GA/10337 April 6 2005

External links

- Homepage of the UN Trusteeship Council (http://www.un.org/en/mainbodies/trusteeship)
- UN Decolonization page (http://www.un.org/Depts/dpi/decolonization/)

Philippines

Republic of the Philippines ***Republika ng Pilipinas***		

Motto: *Maka-Diyos, Maka-Tao, Makakalikasan, at Makabansa*[1] ("For God, People, Nature, and Country")		
Anthem: *Lupang Hinirang*		
 Location of Philippines(green) in ASEAN(dark grey) — [Legend]		
Capital		Manila 14°35′N 121°0′E
Largest city		Quezon City
Official language(s)		Filipino (based on Tagalog) English
Recognised regional languages		Bikol, Cebuano, Hiligaynon, Ilokano, Pampango, Pangasinense, Tagalog, Waray[2]
Optional languages[a]		Spanish and Arabic[3]
National language		Filipino
Demonym		Filipino/Filipina
Government		Unitary presidential constitutional republic
-	President	Benigno Aquino III (LP)
-	Vice President	Jejomar Binay (PDP-Laban)
-	Senate President	Juan Ponce Enrile (PMP)

-	House Speaker	Feliciano Belmonte, Jr. (LP)
-	Supreme Court Chief Justice	Renato Corona
Legislature		Congress
-	Upper House	Senate
-	Lower House	House of Representatives
Independence		from Spain[b] from United States
-	Established	April 27, 1565
-	Declared	June 12, 1898
-	Self-government	March 24, 1934
-	Recognized	July 4, 1946
-	Current constitution	February 2, 1987
Area		
-	Land	299,764 km^2 [2] (72nd) 115,831 sq mi
-	Water (%)	0.61%[4] (inland waters)
Population		
-	2010 estimate	94,013,200 [5] (12th)
-	2007 census	88,574,614[6]
-	Density	306.6/km^2 (43rd) 794.1/sq mi
GDP (PPP)		2010 estimate
-	Total	$351.370 billion[7] (33rd)
-	Per capita	$3,737[7]
GDP (nominal)		2010 estimate
-	Total	$188.719 billion[7] (46th)
-	Per capita	$2,007[7]
Gini (2006)		45.8[4] (medium)
HDI (2010)		▲ 0.638[8] (medium) (97th)
Currency		Peso (Filipino: piso) ₱ (PHP)
Time zone		PST (UTC+8)
-	Summer (DST)	not observed (UTC+8)
Drives on the		right[9]
Internet TLD		.ph
Calling code		+63
1	**a** The 1987 Philippine constitution specifies, "Spanish and Arabic shall be promoted on a voluntary and optional basis."	

2	**b** Philippine revolutionaries declared independence from Spain on June 12, 1898, but the Spanish claim of sovereignty was passed from Spain to the United States in the Treaty of Paris. This led to the Philippine–American War.
3	**c** Rankings above were taken from associated Wikipedia pages as of October 2010 and may be based on data or data sources other than those appearing here.

The **Philippines** i/ˈfɪlɪpiːnz/ (Filipino: *Pilipinas* Tagalog pronunciation: [ˌpɪlɪˈpinɐs]), officially known as the **Republic of the Philippines** (Filipino: *Republika ng Pilipinas*), is a country in Southeast Asia in the western Pacific Ocean. To its north across the Luzon Strait lies Taiwan. West across the South China Sea sits Vietnam. The Sulu Sea to the southwest lies between the country and the island of Borneo, and to the south the Celebes Sea separates it from other islands of Indonesia. It is bounded on the east by the Philippine Sea. Its location on the Pacific Ring of Fire and its tropical climate make the Philippines prone to earthquakes and typhoons but have also endowed the country with natural resources and made it one of the richest areas of biodiversity in the world. An archipelago comprising 7,107 islands, the Philippines is categorized broadly into three main geographical divisions: Luzon, Visayas, and Mindanao. Its capital city is Manila.

With an estimated population of about 94 million people, the Philippines is the world's 12th most populous country. An additional 11 million Filipinos live overseas. Multiple ethnicities and cultures are found throughout the islands. In prehistoric times, Negritos were some of the archipelago's earliest inhabitants. They were followed by successive waves of Austronesian peoples who brought with them influences from Malay, Hindu, and Islamic societies. Trade introduced Chinese cultural influences which remain to this day.

The arrival of Ferdinand Magellan in 1521 marked the beginning of an era of Spanish interest and eventual dominance. In 1543, Spanish explorer Ruy López de Villalobos named the archipelago *Las Islas Filipinas* in honor of Philip II of Spain. Miguel López de Legazpi arrived in the Philippines in 1565 and consolidated Spanish rule in the islands, which remained a colony of Spain for more than 300 years.

Manila became the Asian hub of the Manila–Acapulco galleon fleet. Christianity was widely adopted. As the 19th century gave way to the 20th, there followed in quick succession the Philippine Revolution which spawned the short-lived First Philippine Republic, the Spanish-American War, and the Philippine–American War. In the aftermath, the United States emerged as the dominant power. Aside from the period of Japanese occupation, the United States retained sovereignty over the islands until the end of World War II when the Philippines gained independence. The United States bequeathed to the Philippines the English language and a stronger affinity for Western culture. Since independence the Philippines has had an often tumultuous experience with democracy, with popular "People Power" movements overthrowing a dictatorship in one instance but also underlining the institutional weaknesses of its constitutional republic in others.

Etymology

The name *Philippines* is derived from that of King Philip II of Spain. Spanish explorer Ruy López de Villalobos during his expedition in 1542 named the islands of Leyte and Samar *Felipinas* after the then Prince of Asturias. Eventually the name *Las Islas Filipinas* would be used to cover all the islands of the archipelago. Before it became commonplace, other names such as *Islas del Poniente* (Islands of the West) and Magellan's name for the islands *San Lázaro* were also used by the Spanish to refer to the islands.[10] [11] [12] [13] [14]

The official name of the Philippines has changed several times in the course of the country's history. During the Philippine Revolution, the Malolos Congress proclaimed the establishment of the *República Filipina* or the *Philippine Republic*. From the period of the Spanish-American War and the Philippine–American War until the Commonwealth period, American colonial authorities referred to the country as the *Philippine Islands*, a translation of the Spanish name. During the American period the name *Philippines* began to appear and it has since become the country's common name.[15] Since independence the official name of the country has been the *Republic of the Philippines*.

History

The metatarsal of Callao Man is reported to have been reliably dated by uranium-series dating to 67,000 years ago[16] thereby replacing the Tabon Man of Palawan, carbon-dated to around 24,000 years ago[17] [18] as the oldest human remains found in the archipelago. Negritos were among the archipelago's earliest inhabitants but their appearance in the Philippines has not been reliably dated.[19] There are several opposing theories regarding the origins of ancient Filipinos. F. Landa Jocano theorizes that the ancestors of the Filipinos evolved locally. Wilhelm Solheim's Island Origin Theory[20] postulates that the peopling of the archipelago transpired via trade networks originating in the antediluvian Sundaland area around 48000 to 5000 BCE rather than by wide-scale migration. The Austronesian Expansion Theory states that Malayo-Polynesians coming from Taiwan began migrating to the Philippines around 4000 BCE, displacing earlier arrivals.[21] [22] Whatever the case, by 1000 BCE the inhabitants of the archipelago had developed into four kinds of social groups: hunter-gathering tribes, warrior societies, petty plutocracies, and maritime-centered harbor principalities.[23]

Trade between the maritime-oriented peoples and other Asian countries during the subsequent period brought influences from Hinduism, Buddhism, and Islam. During this time there was no unifying political state encompassing the entire Philippine Archipelago. Instead, the islands were divided among competing thalassocracies ruled by various datus, rajahs, or sultans. These thalassocracies were composed of autonomous barangays which were independent to or allied with larger nations. Among them were the kingdoms of Maynila, Namayan, and Tondo, the confederation of Madyaas, the state of Ma-i, the rajahnates of Butuan and Cebu, and the sultanates of Maguindanao and Sulu.[24] [25] [26] [27] Some of these societies were part of the Malayan empires of Srivijaya, Majapahit, and Brunei.[28] [29] Islam was brought to the Philippines by traders and proselytizers from Malaysia and Indonesia.[30] By the 15th century, Islam was established in the Sulu Archipelago and by 1565 had reached Mindanao, the Visayas, and Luzon.[31]

A page from the Boxer Codex. Left, is a general from the Rajahnate of Butuan and to the right is a princess of Tondo.

In 1521, Portuguese explorer Ferdinand Magellan arrived in the Philippines and claimed the islands for Spain.[32] Colonization began when Spanish explorer Miguel López de Legazpi arrived from Mexico in 1565 and formed the first European settlements in Cebu. In 1571, after dealing with the local royal families in the wake of the Tondo Conspiracy and defeating the Chinese pirate warlord Limahong, the Spanish established Manila as the capital of the Spanish East Indies.[33] [34]

Spanish rule contributed significantly to bringing political unity to the archipelago. From 1565 to 1821, the Philippines was governed as a territory of the Viceroyalty of New Spain and then was administered directly from Madrid after the Mexican War of Independence. The Manila galleons linking Manila to Acapulco traveled once or twice a year between the 16th and 19th centuries. Trade introduced foods such as corn, tomatoes, potatoes, chili peppers, and pineapples from the Americas.[34] Roman Catholic missionaries converted most of the lowland inhabitants to Christianity and founded schools, a university, and hospitals. While a Spanish decree introduced free public schooling in 1863, efforts in mass public education mainly came to fruition during the American period.[35]

During its rule, the Spanish fought off various indigenous revolts and several external colonial challenges from Chinese pirates, the Dutch, and the Portuguese. In an extension of the fighting of the Seven Years' War, British forces under the command of Brigadier General William Draper and Rear-Admiral Samuel Cornish briefly occupied Manila. They found local allies like Diego and Gabriela Silang who took the opportunity to lead a revolt, but Spanish rule was eventually restored following the 1763 Treaty of Paris.[30] [36] [37]

In the 19th century, Philippine ports were opened to world trade and shifts were occurring within Philippine society. Many Spaniards born in the Philippines (*criollos*) and those of mixed ancestry (*mestizos*) became wealthy. The influx of Spanish and Latino settlers secularized churches and opened up government positions traditionally held by Spaniards born in the Iberian Peninsula (*peninsulares*). The ideals of revolution also began to spread through the islands. *Criollo* dissatisfaction resulted in the revolt in Cavite El Viejo in 1872 that was a precursor to the Philippine Revolution.[30] [38] [39] [40] [41]

José Rizal, Marcelo del Pilar, and Mariano Ponce: leaders of the Propaganda Movement

Revolutionary sentiments were stoked in 1872 after three priests—Mariano Gómez, José Burgos, and Jacinto Zamora (collectively known as Gomburza)—were accused of sedition by colonial authorities and executed.[38] [39] This would inspire a propaganda movement in Spain, organized by Marcelo H. del Pilar, José Rizal, and Mariano Ponce, lobbying for political reforms in the Philippines. Rizal was eventually executed on December 30, 1896, on charges of rebellion.[42] As attempts at reform were meeting with resistance, Andrés Bonifacio in 1892 established the secret society called the Katipunan, a society along the lines of the freemasons, which sought independence from Spain through armed revolt.[40] Bonifacio and the Katipunan started the Philippine Revolution in 1896. A faction of the Katipunan, the Magdalo of Cavite province, eventually came to challenge Bonifacio's position as the leader of the revolution and Emilio Aguinaldo took over. In 1898, the Spanish-American War began in Cuba and reached the Philippines. Aguinaldo declared Philippine independence from Spain in Kawit, Cavite on June 12, 1898 and the First Philippine Republic was established the following year. Meanwhile, the islands were ceded by Spain to the United States for US$20 million in the 1898 Treaty of Paris.[43] As it became increasingly clear the United States would not recognize the First Philippine Republic, the Philippine–American War broke out. It ended with American control over the islands which were then administered as an insular area.[44]

In 1935, the Philippines was granted Commonwealth status. Plans for independence over the next decade were interrupted by World War II when the Japanese Empire invaded and established a puppet government. Many atrocities and war crimes were committed during the war such as the Bataan Death March and the Manila massacre that culminated during the Battle of Manila.[45] Allied troops defeated the Japanese in 1945. By the end of the war it is estimated over a million Filipinos had died.[46] On July 4, 1946, the Philippines attained its independence.[4] Immediately after World War II, the Philippines faced a number of challenges. The country had to be rebuilt from the ravages of war. It also had to come to terms with Japanese collaborators. Meanwhile, disgruntled remnants of the Hukbalahap communist rebel army that had previously fought against and resisted the Japanese continued to roam the rural regions. Eventually this threat was dealt with by Secretary of National Defense and later President Ramon Magsaysay but sporadic cases of communist insurgency continued to flare up long afterward.[47] [48] In 1965, Ferdinand Marcos was elected president, his wife Imelda Marcos at his side. Nearing the end of his second term and constitutionally barred from seeking a third, he declared martial law on September 21, 1972. By using political divisions, the tension of the Cold War, and the specter of communist rebellion and Islamic insurgency as justifications, he was able to govern by decree.[49]

A statue of the Virgin Mary built at the EDSA Shrine after the People Power Revolution

On August 21, 1983, Marcos' chief rival opposition leader Benigno "Ninoy" Aquino, Jr. ignored warnings and returned from exile in the United States. He was assassinated as he was taken off the plane at the Manila International Airport (now called the Ninoy Aquino International Airport in his memory). With political pressure building, Marcos eventually called for snap presidential elections in 1986.[47] Corazon Aquino, Benigno's widow, was persuaded to become the presidential candidate and standard bearer of the opposition. The elections were widely thought of as rigged when Marcos was proclaimed the winner. This led to the People Power Revolution, instigated when two long-time Marcos allies—Armed Forces of the Philippines Vice Chief-of-Staff Fidel V. Ramos and Secretary of National Defense Juan Ponce Enrile—resigned and barricaded themselves in Camp Aguinaldo and Camp Crame. Exhorted by the Cardinal Archbishop of Manila Jaime Sin, people gathered in support of the rebel leaders and protested on Epifanio de los Santos Avenue (EDSA). In the face of mass protests and military defections, Marcos and his allies fled to Hawaii and into exile. Corazon Aquino was recognized as president.[48] [50]

The return of democracy and government reforms after the events of 1986 were hampered by national debt, government corruption, coup attempts, a persistent communist insurgency, and Islamic separatists. The economy improved during the administration of Fidel V. Ramos, who was elected president in 1992.[51] However, the economic improvements were negated with the onset of the East Asian financial crisis in 1997. In 2001, amid charges of corruption and a stalled impeachment process, Ramos' successor Joseph Estrada was ousted from the presidency by the 2001 EDSA Revolution and replaced by Gloria Macapagal-Arroyo. As a result of the May 2010 elections, Benigno "Noynoy" Aquino III was elected president.

Politics and government

The Philippines is a constitutional republic with a presidential system of government. It is governed as a unitary state with the exception of the Autonomous Region in Muslim Mindanao which is largely free from the national government. The President functions as both head of state and head of government and is the commander-in-chief of the armed forces. The president is elected by popular vote for a single six-year term, during which he or she appoints and presides over the cabinet.[2]

The bicameral Congress is composed of the Senate, serving as the upper house, with members elected to a six-year term, and the House of Representatives, serving as the lower house, with members elected to a three-year term. The senators are elected at large while the representatives are elected from both legislative districts and through sectoral representation.[2]

The judicial power is vested in the Supreme Court, composed of a Chief Justice as its presiding officer and fourteen associate justices, all of whom are appointed by the President from nominations submitted by the Judicial and Bar Council.[2]

There have been attempts to change the government to a federal, unicameral, or parliamentary government since the Ramos administration.[52] [53]

Security and defense

BRP Gregorio del Pilar, a Hamilton class cutter of the Philippine Navy

Philippine defense is handled by the Armed Forces of the Philippines and is composed of three branches: the Air Force, the Army, and the Navy (including the Marine Corps). Civilian security is handled by Philippine National Police under the Department of the Interior and Local Government (DILG).

In the Autonomous Region in Muslim Mindanao, the largest separatist organization, the Moro National Liberation Front, is now engaging the government politically. Other more militant groups like the Moro Islamic Liberation Front, the communist New People's Army, and the Abu Sayyaf still roam the provinces, but their presence has decreased in recent years due to successful security provided by the Philippine government.[54] [55]

The Philippines has been an ally of the United States since World War II. A mutual defense treaty between the two countries was signed in 1951. The Philippines supported American policies during the Cold War and participated in the Korean and Vietnam wars. It was a member of the now dissolved SEATO, a group that was intended to serve a role similar to NATO and that included Australia, France, New Zealand, Pakistan, Thailand, the United Kingdom, and the United States.[56] After the start of the War on Terror, the Philippines was part of the coalition that gave support to the United States in Iraq.[57] The United States designated the country a major non-NATO ally. The Philippines is currently working to end its domestic insurgency with help from the United States.

International relations

The Philippine Embassy in Washington D.C., United States.

The Philippines' international relations are based on trade with other nations and the well-being of the 11 million overseas Filipinos living outside the country.[58] As a founding and active member of the United Nations, the Philippines has been elected several times into the Security Council. Carlos P. Romulo was a former President of the United Nations General Assembly. The country is an active participant in the Human Rights Council as well as in peacekeeping missions, particularly in East Timor.[59] [60] [61] [62]

In addition to membership in the United Nations, the country is also a founding and active member of ASEAN (Association of Southeast Asian Nations), an organization designed to strengthen relations and promote economic and cultural growth among states in the Southeast Asian region.[63] It has hosted several summits and is an active contributor to the direction and policies of the bloc.[64] The relations it currently enjoys with other Southeast Asian states are in contrast to its relations with them before the 1970s when it was at war with Vietnam and was heavily disputing Sabah with Malaysia; although, disagreements continue to exist due to the Spratly Islands.[65]

The Philippines values its relations with the United States.[58] It supported the United States during the Cold War and the War on Terror and is a major non-NATO ally. Despite this history of goodwill, controversies related to the presence of the now former U.S. military bases in Subic Bay and Clark and the current Visiting Forces Agreement have flared up from time to time.[58] Japan, the biggest contributor of official development assistance to the country,[66] is thought of as a friend. Although historical tensions still exist on issues such as the plight of comfort women much of the animosity inspired by memories of World War II have faded.[67]

Relations with other nations are generally positive. Shared democratic values ease relations with Western and European countries while similar economic concerns help in relations with other developing countries. Historical ties and cultural similarities also serve as a bridge in relations with Spain and Latin America. Despite issues such as domestic abuse and war affecting overseas Filipino workers and obstacles posed by Islamic insurgency in Mindanao, relations with Middle Eastern countries (including Egypt, Iran, Iraq, Libya, Saudi Arabia, and the United Arab Emirates) are friendly as seen in the continuous employment of more than two million overseas Filipinos living there.

With communism no longer the threat it once was, once hostile relations in the 1950s between the Philippines and the People's Republic of China have improved greatly. Issues involving Taiwan, the Spratly Islands, and concerns of expanding Chinese influence, however, still encourage a degree of carefulness.[67] Recent foreign policy has been mostly about economic relations with its Southeast Asian and Asia-Pacific neighbors.[58]

The Philippines is a member of the East Asia Summit (EAS), the Asia-Pacific Economic Cooperation (APEC), the Latin Union, the Group of 24, and the Non-Aligned Movement.[2] It is also seeking to strengthen relations with Islamic countries by campaigning for observer status in the Organisation of Islamic Cooperation.[68] [69]

Administrative divisions

The Philippines is divided into three island groups: Luzon, Visayas, and Mindanao. As of March 2010, these were divided into 17 regions, 80 provinces, 138 cities, 1,496 municipalities, and 42,025 barangays.[70] In addition, Section 2 of Republic Act No. 5446 asserts that the definition of the territorial sea around the Philippine archipelago does not affect the claim over Sabah.[71]

A clickable map of the Philippines exhibiting its 17 regions and 80 provinces.

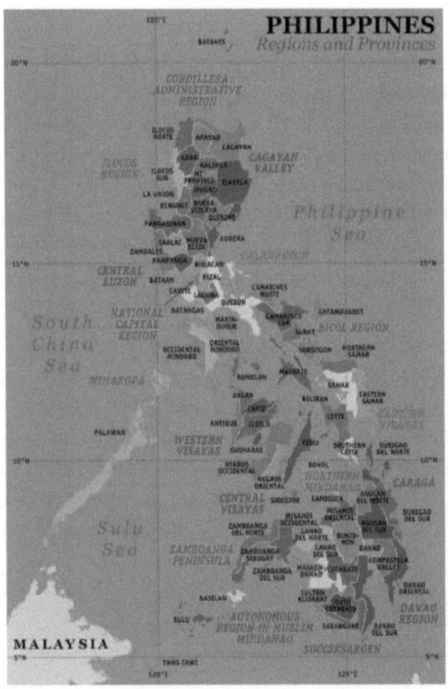

Region	Designation	Regional center
Ilocos Region	Region I	San Fernando, La Union
Cagayan Valley	Region II	Tuguegarao, Cagayan
Central Luzon	Region III	San Fernando, Pampanga
CALABARZON	Region IV-A	Calamba, Laguna
MIMAROPA	Region IV-B	Calapan, Oriental Mindoro
Bicol Region	Region V	Legazpi, Albay
Western Visayas	Region VI	Iloilo City
Central Visayas	Region VII	Cebu City
Eastern Visayas	Region VIII	Tacloban
Zamboanga Peninsula	Region IX	Zamboanga City
Northern Mindanao	Region X	Cagayan de Oro
Davao Region	Region XI	Davao City
SOCCSKSARGEN	Region XII	Koronadal, South Cotabato
Caraga	Region XIII	Butuan
Autonomous Region in Muslim Mindanao	ARMM	Cotabato City
Cordillera Administrative Region	CAR	Baguio
National Capital Region	NCR	Manila

Geography

The Philippines is an archipelago of 7,107 islands[2] with a total land area, including inland bodies of water, of approximately **unknown operator: u','** square kilometers (**unknown operator: u'strong'unknown operator: u','**sq mi). Its **unknown operator: u','** kilometers (**unknown operator: u'strong'unknown operator: u','**mi) of coastline makes it the country with the 5th longest coastline in the world.[2] [72] It is located between 116° 40', and 126° 34' E. longitude and 4° 40' and 21° 10' N. latitude and is bordered by the Philippine Sea to the east, the South China Sea to the west, and the Celebes Sea to the south. The island of Borneo is located a few hundred kilometers southwest and Taiwan is located directly to the north. The Moluccas and Sulawesi are located to the south-southwest and Palau is located to the east of the islands.[2]

Most of the mountainous islands are covered in tropical rainforest and volcanic in origin. The highest mountain is Mount Apo. It measures up to 2,954 meters (9,692 feet) above sea level and is located on the island of Mindanao. The Galathea Depth in the Philippine Trench is the deepest point in the country and the third deepest in the world. The trench is located in the Philippine Sea. The longest river is the Cagayan River in northern Luzon. Manila Bay, upon the shore of which the capital city of Manila lies, is connected to Laguna de Bay, the largest lake in the Philippines, by the Pasig River. Subic Bay, the Davao Gulf, and the Moro Gulf are other important bays. The San Juanico Strait separates the islands of Samar and Leyte but it is traversed by the San Juanico Bridge.[73]

Ancient Filipinos utilized terrace farming to grow crops in the steep mountainous regions of northern Philippines.

Situated on the western fringes of the Pacific Ring of Fire, the Philippines experiences frequent seismic and volcanic activity. The Benham Plateau to the east in the Philippine Sea is an undersea region active in tectonic subduction.[74] Around 20 earthquakes are registered daily, though most are too weak to be felt. The last major earthquake was the 1990 Luzon earthquake.[75] There are many active volcanoes such as the Mayon Volcano, Mount Pinatubo, and Taal Volcano. The eruption of Mount Pinatubo in June 1991 produced the second largest terrestrial eruption of the 20th century.[76] Not all notable geographic features are so violent or destructive. A more serene legacy of the geological disturbances is the Puerto Princesa Subterranean River. The white sand beaches that make Boracay a popular vacation getaway are made of coral remnants.

Due to the volcanic nature of the islands, mineral deposits are abundant. The country is estimated to have the second-largest gold deposits after South Africa and one of the largest copper deposits in the world.[77] It is also rich in nickel, chromite, and zinc. Despite this, poor management, high population density, and environmental consciousness have resulted in these mineral resources remaining largely untapped.[77] Geothermal energy, however, is another product of volcanic activity that the country has harnessed more successfully. The Philippines is the world's second-biggest geothermal producer behind the United States, with 18% of the country's electricity needs being met by geothermal power.[78]

Flora and fauna

The Philippines' rainforests and its extensive coastlines make it home to a diverse range of birds, plants, animals, and sea creatures.[79] It is one of the ten most biologically mega-diverse countries and is at or near the top in terms of biodiversity per unit area.[80] [81] [82] Around 1,100 land vertebrate species can be found in the Philippines including over 100 mammal species and 170 bird species not thought to exist elsewhere.[83] Endemic species include the tamaraw of Mindoro, the Visayan spotted deer, the Philippine mouse deer, the Visayan warty pig, the Philippine flying lemur, and several species of bats.[84]

Philippine tarsier (*Tarsius syrichta*), one of the smallest primates.

The Philippines lacks large predators, with the exception of snakes, such as pythons and cobras, and birds of prey, such as the national bird, known as the Philippine eagle.[85] Other native animals include the palm civet cat, the dugong, and the Philippine tarsier associated with Bohol. With an estimated 13,500 plant species in the country, 3,200 of which are unique to the islands,[83] Philippine rainforests boast an array of flora, including many rare types of orchids and rafflesia.[86] [87] The narra is considered as the most important type of hardwood.[88]

Philippine maritime waters encompass as much as 2.2 million square kilometers (850,000 square miles) producing unique and diverse marine life and is an important part of the Coral Triangle.[71] There are 2,400 fish species and over 500 species of coral.[79] [83] The Apo Reef is the country's largest contiguous coral reef system and the second-largest in the world.[89] Philippine waters also sustain the cultivation of pearls, crabs, and seaweeds.[79] [90]

Deforestation, often the result of illegal logging, is an acute problem in the Philippines. Forest cover declined from 70% of the country's total land area in 1900 to about 18.3% in 1999.[91] [92] Many species are endangered and scientists say that Southeast Asia, which the Philippines is part of, faces a catastrophic extinction rate of 20% by the end of the century.[93] According to Conservation International, "the country is one of the few nations that is, in its entirety, both a hotspot and a megadiversity country, placing it among the top priority hotspots for global conservation."[86]

Climate

Typhoon Megi (Juan) over the Philippines.

The Philippines has a tropical maritime climate and is usually hot and humid. There are three seasons: *tag-init* or *tag-araw*, the hot dry season or summer from March to May; *tag-ulan*, the rainy season from June to November; and *tag-lamig*, the cool dry season from December to February. The southwest monsoon (from May to October) is known as the Habagat, and the dry winds of the northeast monsoon (from November to April), the Amihan.[94] Temperatures usually range from 21°C (70°F) to 32°C (90°F) although it can get cooler or hotter depending on the season. The coolest month is January; the warmest is May.[2] [95]

The average yearly temperature is around 26.6°C (79.88°F).[94] In considering temperature, location in terms of latitude and longitude is not a significant factor. Whether in the extreme north, south, east, or west of the country, temperatures at sea level tend to be in the same range. Altitude usually has more of an impact. The average annual temperature of Baguio at an elevation of 1,500 meters (4,900 feet) above sea level is 18.3°C (64.9°F), making it a popular destination during hot summers.[94] Likewise, Tagaytay is a favored retreat.

Sitting astride the typhoon belt, most of the islands experience annual torrential rains and thunderstorms from July to October,[96] with around nineteen typhoons entering the Philippine area of responsibility in a typical year and eight or nine making landfall.[97] [98] [99] Annual rainfall measures as much as 5,000 millimeters (200 inches) in the mountainous east coast section but less than 1,000 millimeters (39 inches) in some of the sheltered valleys.[96] The wettest known tropical cyclone to impact the archipelago was the July 1911 cyclone, which dropped over 1168 millimetres (46.0 in) of rainfall within a 24-hour period in Baguio City.[100] *Bagyo* is the local term for a tropical cyclone in the Philippines.[100]

Economy

The national economy of the Philippines is the 46th largest in the world, with an estimated 2010 gross domestic product (nominal) of $189 billion.[7] Primary exports include semiconductors and electronic products, transport equipment, garments, copper products, petroleum products, coconut oil, and fruits.[4] Major trading partners include the United States, Japan, China, Singapore, South Korea, the Netherlands, Hong Kong, Germany, Taiwan, and Thailand.[4] Its unit of currency is the Philippine peso (₱ or PHP).

Makati, the country's financial center.

A newly industrialized country, the Philippine economy has been transitioning from one based on agriculture to one based more on services and manufacturing. Of the country's total labor force of around 38.1 million,[4] the agricultural sector employs close to 32% but contributes to only about 13.8% of GDP. The industrial sector employs around 13.7% of the workforce and accounts for 30% of GDP. Meanwhile the 46.5% of workers involved in the services sector are responsible for 56.2% of GDP.[101] [102]

The unemployment rate as of July 2009 stands at around 7.6% and due to the global economic slowdown inflation as of September 2009 reads 0.70%.[102] Gross international reserves as of July 2011 are $75.174 billion.[103] In 2004, public debt as a percentage of GDP was estimated to be 74.2%; in 2008, 56.9%.[4] Gross external debt has risen to $66.27 billion.[4] The country is a net importer.[102]

The Philippine Stock Exchange with the statue of Benigno Aquino, Jr.

After World War II, the country was for a time regarded as the second wealthiest in East Asia, next only to Japan.[58] [104] [105] However, by the 1960s its economic performance started being overtaken. The economy stagnated under the dictatorship of Ferdinand Marcos as the regime spawned economic mismanagement and political volatility.[58] [105] The country suffered from slow economic growth and bouts of economic recession. Only in the 1990s with a program of economic liberalization did the economy begin to recover.[58] [105]

The 1997 Asian Financial Crisis affected the economy, resulting in a lingering decline of the value of the peso and falls in the stock market. But the extent it was affected initially was not as severe as that of some of its Asian neighbors. This was largely due to the fiscal conservatism of the government, partly as a result of decades of monitoring and fiscal supervision from the International Monetary Fund (IMF), in comparison to the massive spending of its neighbors on the rapid acceleration of economic growth.[51] There have been signs of progress since. In 2004, the economy experienced 6.4% GDP growth and 7.1% in 2007, its fastest pace of growth in three decades.[7] [106] [107] Yet average annual GDP growth per capita for the period 1966–2007 still stands at 1.45% in comparison to an average of 5.96% for the East Asia and the Pacific region as a whole and the daily income for 45% of the population of the Philippines remains less than $2.[108] [109] Despite enjoying sustained economic growth during the first decade of the 21st century, as of 2010, the country's economy remains smaller than those of its Southeast Asian neighbors Indonesia, Thailand, Malaysia, and Singapore in terms of GDP and GDP per capita (nominal).[110]

Other incongruities and challenges exist. The economy is heavily reliant on remittances which surpass foreign direct investment as a source of foreign currency. Regional development is uneven with Luzon—Metro Manila in particular—gaining most of the new economic growth at the expense of the other regions,[111] although the government has taken steps to distribute economic growth by promoting investment in other areas of the country. Despite constraints, service industries such as tourism and business process outsourcing have been identified as areas with some of the best opportunities for growth for the country.[102] [112] Goldman Sachs includes the country in its list of the "Next Eleven" economies.[113] But China and India have emerged as major economic competitors.[114]

The Philippines is a member of the World Bank, the International Monetary Fund, the World Trade Organization (WTO), the Asian Development Bank which is headquartered in Mandaluyong City, the Colombo Plan, and the G-77 among other groups and institutions.[4]

Demographics

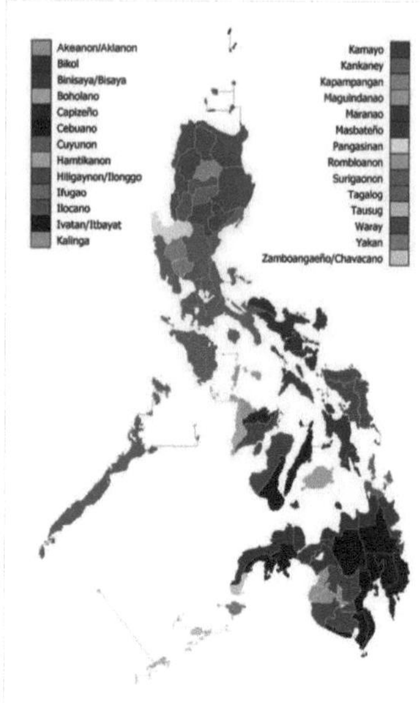

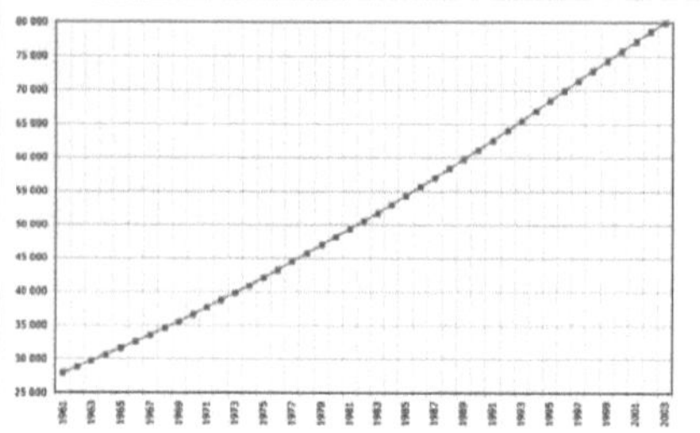
Population growth of the Philippines.

Population in Philippines[115]	
Year	**Million**
1971	37.6
1980	48.1
1990	62.4
2000	77.7
2004	83.9
2008	90.3
Source: OECD/World Bank	

Population in Philippines increased from 1990 to 2008 with 28 million with 45 % growth.[115] The first official census in the Philippines was carried out in 1877 and recorded a population of 5,567,685.[116] As of 2011, the Philippines has become the world's 12th most populous nation, with a population of over 94 million. It is estimated that half of the population resides on the island of Luzon. The population growth rate between 1995 to 2000 of 3.21% decreased to an estimated 1.95% for the 2005 to 2010 period, but remains a contentious issue.[6] [117] The population's median age is 22.7 years with 60.9% aged from 15 to 64 years old.[4] Life expectancy at birth is 71.38 years, 74.45 years for females and 68.45 years for males.[118]

There are about 11 million Filipinos outside the Philippines.[119] Since the liberalization of United States immigration laws in 1965, the number of people in the United States having Filipino ancestry has grown substantially. In 2007 there were an estimated 3.1 million.[120] [121] According to the United States Census Bureau, immigrants from the Philippines made up the second largest group after Mexico that sought family reunification.[122] Some two million Filipinos work in the Middle East, with nearly a million in Saudi Arabia alone.[123]

Ethnicity

According to the 2000 census 28.1% of Filipinos are Tagalog, 13.1% Cebuano, 9% Ilocano, 7.6% Bisaya/Binisaya, 7.5% Hiligaynon, 6% Bikol, 3.4% Waray, and 25.3% are classified as other.[4] [124] These general headings can be broken down further to yield more distinct non-tribal groups like the Moro, the Kapampangan, the Pangasinense, the Ibanag, and the Ivatan.[125] There are also indigenous peoples like the Igorot, the Lumad, the Mangyan, the Bajau, and the tribes of Palawan.[126] Negritos, such as the Aeta and the Ati, are considered among the earliest inhabitants of the islands.[127]

Filipinos generally belong to several Asian ethnic groups classified linguistically as part of the Austronesian or Malayo-Polynesian speaking people.[126] It's believed that thousands of years ago Taiwanese aborigines migrated to the Philippines from Taiwan, bringing with them knowledge of agriculture and ocean-sailing, and displacing the earlier Negrito groups of the islands.[128]

Eventually Chinese, Spanish, and American arrivals intermarried with the various indigenous ethnic groups that had evolved.[128] Their descendants are known as mestizos.[129] Chinese Filipinos number about two million.[130] Other migrant ethnic groups who have settled in the country from elsewhere include Arabs, Britons, other Europeans, Indonesians, Japanese, Koreans, and South Asians.

Cities

Metro Manila is the most populous of the twelve defined metropolitan areas in the Philippines and the 11th most populous in the world. As of the 2007 census, it had a population of 11,553,427, comprising 13% of the national population.[131] Including suburbs in the adjacent provinces (Bulacan, Cavite, Laguna, and Rizal) of Greater Manila, the population is around 21 million.[131] [132]

Metro Manila's gross regional product is estimated as of July 2009 to be ₱468.4 billion (at constant 1985 prices) and accounts for 33% of the nation's GDP.[133] In 2008, it ranked as the 40th wealthiest urban agglomeration in the world and the 2nd in Southeast Asia, according to PricewaterhouseCoopers.[134] Davao City in Mindanao and Cebu City in the Visayas are other important urban centers.

Language

Native Languages (2000)[135]	
Tagalog	22 million
Cebuano	20 million
Ilokano	7.7 million
Hiligaynon	7 million
Waray-Waray	3.1 million

Ethnologue lists 175 individual languages in the Philippines, 171 of which are living languages while 4 no longer have any known speakers. They are part of the Borneo–Philippines group of the Malayo-Polynesian languages, which is itself a branch of the Austronesian language family.[126]

According to the 1987 Philippine Constitution, Filipino and English are the official languages. Filipino is a de facto version of Tagalog, spoken mainly in Metro Manila and other urban regions. Both Filipino and English are used in government, education, print, broadcast media, and business. The constitution designates regional languages such as Bicolano, Cebuano, Ilocano, Hiligaynon, Kapampangan, Pangasinan, Tagalog, and Waray-Waray as auxiliary official languages, and mandates that Spanish and Arabic shall be promoted on a voluntary and optional basis.[3]

Other languages such as Aklanon, Boholano, Chavacano, Zamboangueño, Cuyonon, Ifugao, Itbayat, Ivatan, Kalinga, Kamayo, Kankana-ey, Kinaray-a, Maguindanao, Maranao, Masbatenyo, Romblomanon, Surigaonon, Tausug,

Yakan, and several Visayan languages are prevalent in their respective provinces.

Religion

More than 90% of the population are Christians: about 80% belong to the Roman Catholic Church while 10% belong to other Christian denominations, such as the Iglesia ni Cristo, the Philippine Independent Church, the Seventh-day Adventist Church, United Church of Christ in the Philippines, and Jehovah's Witnesses.[136] The Philippines is one of two predominantly Roman Catholic countries in Asia, the other being East Timor.

Between five and ten percent of the population are Muslim, most of whom live in parts of Mindanao, Palawan, and the Sulu Archipelago—an area known as Bangsamoro or the Moro region.[137] [138] Some have migrated into urban and rural areas in different parts of the country. Most Muslim Filipinos practice Sunni Islam according to the Shafi'i school.[31] Philippine traditional religions are still practiced by many aboriginal and tribal groups, often syncretized with Christianity and Islam. Animism, folk religion, and shamanism remain present as undercurrents of mainstream religion, through the *albularyo*, the *babaylan*, and the *manghihilot*. Buddhism, Taoism, and Chinese folk religion, are dominant in Chinese communities.[138] There are also followers of Judaism and Baha'i.[139]

Education

The National Statistics Office reports a simple literacy rate of 93.4% and a functional literacy rate of 84.1% for 2003.[4] [102] [108] Literacy is about equal for males and females.[4] Spending for education is around 2.5% of GDP.[4] According to the Department of Education, or DepEd, there were 44,846 elementary schools and 10,384 secondary schools registered for the school year 2009–2010[140] while the Commission on Higher Education (CHED) lists 2,180 higher education institutions, 607 of which are public and 1,573 private.[141] Classes start in June and end in March. The majority of colleges and universities follow a semester calendar from June to October and November to March. There are a number of foreign schools with study programs.[2] Republic Act No. 9155 gives the framework of basic education in the Philippines and provides for compulsory elementary education and free high school education.[142]

Several government agencies are involved with education. The Department of Education covers elementary, secondary, and nonformal education; the Technical Education and Skills Development Authority (TESDA) administers the post-secondary middle-level education training and development; and the Commission on Higher Education (CHED) supervises the college and graduate academic programs and degrees as well as regulates standards in higher education.[143] In 2004, madrasahs were mainstreamed in 16 regions nationwide mainly in Muslim areas in Mindanao under the auspices and program of the Department of Education.[144] Public universities are all non-sectarian entities, and are further classified as State University and College (SUC) or Local College and University (LCU).[141] SUCs are funded by the national government as determined by the Philippine Congress.[145] The University of the Philippines is the national university of the Philippines.[146]

Health

Most of the national burden of health care is taken up by private health providers. In 2006, total expenditures on health represented 3.8% of GDP. 67.1% of that came from private expenditures while 32.9% was from government. External resources accounted for 2.9% of the total. Health expenditures represented about 6.1% of total government spending. Per capita total expenditure at average exchange rate was $52.[147] The proposed national health budget for 2010 is ₱28 billion (about $597 million) or ₱310 ($7) per person.[148] The government share of total spending on health has declined steadily, and with more people, there has been less to spend per person.

There are an estimated 90,370 physicians or 1 per every 833 people, 480,910 nurses, 43,220 dentists, and 1 hospital bed per every 769 people.[147] Retention of skilled practitioners is a problem. 70% of nursing graduates go overseas to work. The country is the biggest supplier of nurses.[149] In 2001 there were about 1,700 hospitals, of which about

40% were government-run and 60% private. Cardiovascular diseases account for more than 25% of all deaths. According to official estimates, 1,965 cases of human immunodeficiency virus (HIV) were reported in 2003, of which 636 had developed acquired immune deficiency syndrome (AIDS). Other estimates have as many as 12,000 people living with HIV/AIDS in 2005.[150]

Infrastructure

Transportation

A jeepney and a bus, common forms of public transport in the Philippines

The transportation infrastructure in the country is relatively underdeveloped. Partly this is due to the mountainous terrain and the scattered geography of the islands, but it is also the result of the government's persistent underinvestment in infrastructure. In 2003, only 3.6% of GDP went to infrastructure development which was significantly lower than that of some of its neighbors.[96] Consequently, while there are **unknown operator: u','** kilometers (**unknown operator: u'strong'unknown operator: u','**mi) of roads in the country, only around 20% of the total is paved.[151]

Nevertheless there are many ways to get around, especially in urban areas. Buses, jeepneys, taxis, and motorized tricycles are commonly available in major cities and towns. In 2007, there were about 5.53 million registered motor vehicles with registration increasing at an average annual rate of 4.55%.[152] Train services are provided by three main railway networks that serve different areas of Metro Manila and parts of Luzon: the Manila Light Rail Transit System (LRT), the Manila Metro Rail Transit System (MRT), and the Philippine National Railways (PNR).

Terminal 3 of Ninoy Aquino International Airport, the main gateway to the country

As an archipelago, inter-island travel via watercraft is often necessary. The busiest seaports are Manila, Cebu, Iloilo, Davao, Cagayan de Oro, and Zamboanga.[153] Passenger ships and other sea vessels such as those operated by Superferry, Negros Navigation, and Sulpicio Lines serve Manila, with links to various cities and towns. In 2003, the 919-kilometer (571 mi) Strong Republic Nautical Highway (SRNH), an integrated set of highway segments and ferry routes covering 17 cities was established.[154]

Some rivers that pass through metropolitan areas, such as the Pasig River and Marikina River, have air-conditioned commuter ferries. The Pasig River Ferry Service has numerous stops in Manila, Makati, Mandaluyong, Pasig and Marikina.[155] There are **unknown operator: u','** kilometers (**unknown operator: u'strong'unknown operator: u','**mi) of navigable inland waterways.[4]

There are 85 public airports in the country, and around 111 more that are private.[151] The Ninoy Aquino International Airport (NAIA) is the main international airport. Other important airports include the Diosdado Macapagal International Airport, Mactan-Cebu International Airport, Francisco Bangoy International Airport and Zamboanga International Airport. Philippine Airlines, Asia's oldest commercial airline still operating under its original name, and Cebu Pacific, the leading low-cost airline, are the major airlines serving most domestic and international destinations.[156] [157] [158]

Communications

The Philippines has a sophisticated cellular phone industry and a high concentration of users.[159] As of 2008, there are about 67.9 million cellular phone subscribers in the Philippines.[160] Text messaging is a popular form of communication and has fostered a culture of quick greetings and forwarded jokes among Filipinos. In 2007, the nation sent an average of one billion SMS messages per day.[161] Out of this growing number of avid text message senders, over five million of them use their cellular phones as virtual wallets, making it a leader among developing nations in providing financial transactions over cellular networks.[162]

The Philippine Long Distance Telephone Company commonly known as PLDT is the leading telecommunications provider. It is also the largest company in the country.[159] [163] Its wholly owned subsidiaries Smart Communications and Piltel, along with Globe Telecom of the Ayala Group, BayanTel, and Sun Cellular are the major cellular service providers in the country.

There are approximately 383 AM and 659 FM radio stations and 297 television and 873 cable television stations.[164] Estimates for internet penetration in the Philippines vary widely ranging from a low of 2.5 million to a high of 24 million people.[165] [166] Social networking and watching videos are among the most frequent internet activities.[167] [168]

Culture and society

Philippine culture is a combination of Eastern and Western cultures. The Philippines exhibits aspects found in other Asian countries with a Malay[169] heritage, yet its culture also displays a significant amount of Spanish and American influences. Traditional festivities known as *barrio fiestas* (district festivals) to commemorate the feast days of patron saints are common. The Moriones Festival and Sinulog Festival are a couple of the most well-known. These community celebrations are times for feasting, music, and dancing. Some traditions, however, are changing or gradually being forgotten due to modernization. The Bayanihan Philippine National Folk Dance Company has been lauded for preserving many of the various traditional folk dances found throughout the Philippines. They are famed for their iconic performances of Philippine dances such as the *tinikling* and *singkil* that both feature the use of clashing bamboo poles.[170]

Percussion instruments that make up the Philippine kulintang ensemble, an example of pre-Hispanic musical tradition

Barasoain Church in Malolos, Bulacan where the First Philippine Republic was founded.

One of the most visible Hispanic legacies is the prevalence of Spanish names and surnames among Filipinos. However, a Spanish name and surname does not necessarily denote Spanish ancestry. This peculiarity, unique among the people of Asia, came as a result of a colonial decree, the Clavería edict, for the systematic distribution of family names and implementation of the Spanish naming system on the population.[171] The names of many streets, towns, and provinces are also in Spanish. Spanish architecture has left an imprint in the Philippines in the way many towns were designed around a central square or *plaza mayor*, but many of the buildings bearing its influence were demolished during World War II.[24] Some examples remain, mainly among the country's churches, government buildings, and universities. Four Philippine baroque churches are included in the list of UNESCO World Heritage Sites: the San Agustín Church in Manila, the Paoay Church in Ilocos Norte, the Nuestra Señora de la Asunción (Santa María)

Church in Ilocos Sur, and the Santo Tomás de Villanueva Church in Iloilo.[172] Vigan in Ilocos Sur is also known for the many Hispanic-styled houses and buildings preserved there.[173]

The common use of the English language is an example of the American impact on Philippine society. It has contributed to the ready acceptance and influence of American pop cultural trends. This affinity is seen in Filipinos' love of fast food, film, and music. Fast food outlets are found on many street corners. American global fast food chain stalwarts have entered the market, but local fast food chains like Goldilocks and most notably Jollibee, the leading fast food chain in the country, have emerged and compete successfully against their foreign rivals.[174] [175] Filipinos regularly listen to and watch contemporary American, Asian, and European music and film just as they enjoy Original Pilipino Music (also known as OPM) and local films.

Cuisine

Philippine cuisine has evolved over several centuries from its Malayo-Polynesian origins to become a mixed cuisine with many Hispanic, Chinese, American, and other Asian influences that have been adapted to local ingredients and the Filipino palate to create distinctively Filipino dishes. Dishes range from the very simple, like a meal of fried salted fish and rice, to the elaborate, such as the *paellas* and *cocidos* created for fiestas. Popular dishes include *lechón*, *adobo*, *sinigang*, *kare-kare*, *tapa*, crispy *pata*, *pancit*, *lumpia*, and *halo-halo*. Some common local ingredients used in cooking are calamondins, coconuts, *saba* (a kind of short wide plantain), mangoes, milkfish, and fish sauce. Filipino taste buds tend to favor robust flavors but the cuisine is not as spicy as those of its neighbors.[175] [176]

The halo-halo is a dessert made of ice, milk, various fruits, and ice cream.

Unlike many of their Asian counterparts, Filipinos do not eat with chopsticks; they use western cutlery. However, possibly due to rice being the primary staple food and the popularity of a large number of stews and main dishes with broth in Philippine cuisine, the main pairing of utensils seen at the Filipino dining table is that of spoon and fork, not knife and fork.[177] The traditional way of eating with the hands known as *kamayan* is seen more often in less urbanized areas.[178]

Mythology and literature

Philippine mythology has been handed down primarily through the traditional oral folk literature of the Filipino people. While each unique ethnic group has its own stories and myths to tell, Hindu and Spanish influence can nonetheless be detected in many cases. Many of the myths are creation stories or stories about supernatural creatures, such as the *aswang* (vampire), the *diwata* (fairy), and Nature. Some popular figures from Philippine mythologies are Maria Makiling, Lam-Ang, and the Sarimanok.[179]

Philippine literature comprises works usually written in Filipino, Spanish, or English. Some of the most known were created in the 19th century. Francisco Balagtas the poet and playwright who wrote *Florante at Laura* is recognized as a preeminent writer in the Filipino language. José Rizal wrote the novels *Noli Me Tangere* (*Touch Me Not*) and *El Filibusterismo* (*The Filibustering*, also known as *The Reign of Greed*) and is considered a national hero. His depiction of the injustices of Spanish rule, and his death by firing squad, inspired other Philippine revolutionaries to seek independence. In the 20th century, among those officially recognized as National Artists of the Philippines in literature are N.V.M. Gonzalez, Nick Joaquin, F. Sionil Jose, and Alejandro Roces.[180]

Media

Philippine media uses mainly Filipino and English. Other Philippine languages, including various Visayan languages are also used, especially in radio due to its ability to reach remote rural locations that might otherwise not be serviced by other kinds of media. The dominant television networks ABS-CBN, GMA and TV5 also have extensive radio presence.[181]

The entertainment industry is vibrant and feeds broadsheets and tabloids with an unending supply of details about celebrities and sensationalist scandals du jour. Drama and fantasy shows are anticipated as are Latin telenovelas, Asianovelas, and anime. Daytime television is dominated by game shows, variety shows, and talk shows such as *Eat Bulaga*, *Showtime*, and *Happy, Yipee, Yehey*.[182] Philippine cinema has a long history and is popular domestically, but has faced increasing competition from American, Asian and European films. Critically acclaimed directors and actors include Lino Brocka and Nora Aunor for films like *Maynila: Sa mga Kuko ng Liwanag* (*Manila: In the Claws of Light*) and *Himala* (*Miracle*). In recent years it has become common to see celebrities flitting between television and movies and then moving into politics provoking concerns.[183]

Sports and recreation

Various sports and pastimes are popular in the Philippines including basketball, boxing, volleyball, football, badminton, karate, taekwondo, billiards, ten-pin bowling, chess, and sipa. motocross, cycling, and mountaineering are also becoming popular. Basketball is played at both amateur and professional levels and is considered to be the most popular sport in the Philippines.[184] In almost every corner of the cities, there is a basketball court.[170]

A PBA basketball game at the Smart Araneta Coliseum, Southeast Asia's largest arena.

Traditional Filipino games such as *luksung baka*, *patintero*, *piko*, and *tumbang preso* are still played primarily as children's games among the youth.[185] [186] *Sungka* is a traditional native Filipino board game. Card games are popular during festivities, with some, including *pusoy* and *tong-its*, being used as a form of illegal gambling. Mahjong is played in some Filipino communities. The yo-yo, a popular toy in the Philippines, was introduced in its modern form by Pedro Flores with its name from the Ilokano language.[187]

Arnis (Eskrima or Kali in some regions) is the national martial art and sport.[188] Today there are said to be almost as many Filipino fighting styles as there are islands in the Philippines. In 1972, the Philippine government included Filipino martial arts into the national sports arena. The Ministry of Education, Culture and Sports also incorporated them into the physical education curriculum for high school and college students.

Some Filipinos recognized for their achievements include Francisco Guilledo, Flash Elorde, and Manny Pacquiao in boxing; Paulino Alcántara in football (soccer); Carlos Loyzaga, Robert Jaworski, and Ramon Fernandez in basketball; Efren Reyes in billiards; Eugene Torre in chess; and Rafael Nepomuceno in bowling.[189] [190] [191]

See also

- International rankings of the Philippines

References

[1] "Republic Act No. 8491" (http://web.archive.org/web/20071205235342/http://www.gov.ph/aboutphil/RA8491.asp). Republic of the Philippines. Archived from the original (http://www.gov.ph/aboutphil/RA8491.asp) on 2007-12-05. . Retrieved 2008-09-30. was revisited on November 19, 2010

[2] General Information (http://web.archive.org/web/20071022221129/http://www.gov.ph/aboutphil/general.asp). (older version -- as it existed in 2007 -- during the presidency of Gloria Macapagal Arroyo), *The Official Government Portal of the Republic of the Phllippines* (http://www.gov.ph/).

[3] 1987 Constitution of the Republic of the Philippines, Article XIV, Section 7. (http://www.chanrobles.com/article14language.htm) Retrieved 2009-11-21 from the Chan Robles Virtual Law Library (http://www.chanrobles.com/).

[4] Central Intelligence Agency. (2009-10-28). "East & Southeast Asia :: Philippines" (https://www.cia.gov/library/publications/the-world-factbook/geos/rp.html). *The World Factbook.* Washington, DC: Author. . Retrieved 2009-11-07.

[5] Department of Economic and Social Affairs Population Division (2010). *National Statistics Office 2000 Census-based Population Projection* (http://www.census.gov.ph/data/sectordata/popproj_tab1r.html). 2010 revision. Philippine National Statistics Office. . Retrieved 2011-05-21.

[6] Republic of the Philippines. National Statistics Office. (2008). "Official population count reveals.." (http://www.census.gov.ph/data/pressrelease/2008/pr0830tx.html). Author. . Retrieved 2008-04-17.

[7] "Philippines" (http://www.imf.org/external/pubs/ft/weo/2011/01/weodata/weorept.aspx?sy=2008&ey=2011&scsm=1&ssd=1&sort=country&ds=.&br=1&c=566&s=NGDPD,NGDPDPC,PPPGDP,PPPPC,LP&grp=0&a=&pr.x=35&pr.y=15). International Monetary Fund. . Retrieved 2011-05-06.

[8] United Nations Development Programme. (2010). Table 1 – Human Development Index and its components (http://hdr.undp.org/en/media/HDR_2010_EN_Table1.pdf). *Human Development Report 2010 – The Real Wealth of Nations: Pathways to Human Development.* Palgrave Macmillan. ISBN 9780230284456 90101.

[9] Lucas, Brian. (August 2005). "Which side of the road do they drive on?" (http://www.brianlucas.ca/roadside/). . Retrieved 2009-02-22.

[10] Scott, William Henry. (1994). *Barangay: Sixteenth-century Philippine Culture and Society* (http://books.google.com/?id=15KZU-yMuisC). Ateneo de Manila University Press. p. 6 (http://books.google.com/books?id=15KZU-yMuisC&pg=PA6). ISBN 9715501354. . Retrieved 14 May 2009.

[11] Spate, Oskar H. K. (1979). "Chapter 4. Magellan's Successors: Loaysa to Urdaneta. Two failures: Grijalva and Villalobos" (http://epress.anu.edu.au/spanish_lake/mobile_devices/ch04s05.html). *The Spanish Lake – The Pacific since Magellan, Volume I* (http://epress.anu.edu.au/spanish_lake/mobile_devices/). Taylor & Francis. p. 97. ISBN 070990049X. . Retrieved 2010-01-07.

[12] Friis, Herman Ralph. (Ed.). (1967). *The Pacific Basin: A History of Its Geographical Exploration* (http://books.google.com/?id=veuwAAAAIAAJ&cd=5&dq=islas+del+poniente+san+lazaro&q=islas+del+poniente#search_anchor). American Geographical Society. p. 369. . Retrieved 2010-01-07.

[13] Galang, Zoilo M. (Ed.). (1957). *Encyclopedia of the Philippines, Volume 15* (http://books.google.com/?id=lt5uAAAAMAAJ&cd=2&dq=islas+del+poniente+san+lazaro&q=islas+del+poniente+#search_anchor) (3rd ed.). E.Floro. p. 46. . Retrieved 2010-01-07.

[14] Tarling, Nicholas. (1999). *The Cambridge History of Southeast Asia – Volume One, Part Two – From c. 1500 to c. 1800* (http://books.google.com/?id=jtsMLNmMzbkC&printsec=frontcover#v=onepage&q). Cambridge, UK: Cambridge University Press. p. 12 (http://books.google.com/books?id=jtsMLNmMzbkC&pg=PA12&dq=#v=onepage&q&f=false). ISBN 0521663709. . Retrieved 2010-01-07.

[15] Quezon, Manuel, III. (2005-03-28). "The Philippines *are* or *is*?" (http://www.quezon.ph/2005/03/28/323/). *Manuel L. Quezon III: The Daily Dose.* Retrieved 2009-12-20.

[16] Henderson, Barney (August 4, 2010). "Archaeologists unearth 67000-year-old human bone in Philippines" (http://www.telegraph.co.uk/news/worldnews/asia/philippines/7924538/Archaeologists-unearth-67000-year-old-human-bone-in-Philippines.html). *The Daily Telegraph* (London). . Retrieved August 4, 2010.

[17] Fox, Robert B. (1970). *The Tabon Caves: Archaeological Explorations and Excavations on Palawan* (http://books.google.com/?id=pd6AAAAAMAAJ&q=tabon+man). National Museum. p. 44. ASIN B001O7GGNI. . Retrieved 2009-12-16.

[18] Scott, William Henry. (1984). *Prehispanic Source Materials for the Study of Philippine History* (http://books.google.com/?id=FSlwAAAAMAAJ&q=pre-mongoloid). Quezon City: New Day Publishers. p. 15. ISBN 9711002272. .

[19] Scott, William Henry. (1984). *Prehispanic Source Materials for the Study of Philippine History* (http://books.google.com/?id=FSlwAAAAMAAJ&q=pygmy+Negrito). Quezon City: New Day Publishers. p. 138. ISBN 9711002272. . "Not one roof beam, not one grain of rice, not one pygmy Negrito bone has been recovered. Any theory which describes such details is therefore pure hypothesis and should be honestly presented as such."

[20] Solheim, Wilhelm G., II. (2006). *Archeology and Culture in Southeast Asia.* University of the Philippines Press. pp. 57–139 (http://books.google.com/books?id=SiOgq0OPT_MC&pg=PA57&lpg=). ISBN 9789715425087.

[21] Solheim, Wilhelm G., II. (January 2006). *Origins of the Filipinos and Their Languages* (http://web.archive.org/web/20080803020434/http://web.kssp.upd.edu.ph/linguistics/plc2006/papers/FullPapers/I-2_Solheim.pdf). Archived from the original (http://web.kssp.upd.

edu.ph/linguistics/plc2006/papers/FullPapers/I-2_Solheim.pdf) on 2008-08-03. . Retrieved 2009-08-27.

[22] Mijares, Armand Salvador B. (2006). The Early Austronesian Migration To Luzon: Perspectives From The Peñablanca Cave Sites (http://ejournal.anu.edu.au/index.php/bippa/article/viewFile/10/9). *Bulletin of the Indo-Pacific Prehistory Association* **26**: 72–78.

[23] Legarda, Benito, Jr. (2001). "Cultural Landmarks and their Interactions with Economic Factors in the Second Millennium in the Philippines". *Kinaadman (Wisdom) A Journal of the Southern Philippines* **23**: 40.

[24] Ring, Trudy, Robert M. Salkin, and Sharon La Boda. (1996). *International Dictionary of Historic Places: Asia and Oceania* (http://books.google.com/?id=vWLRxJEU49EC&pg=PA565&dq=#v=onepage&q=). Taylor & Francis. pp. 565–569. ISBN 1884964044. . Retrieved 2010-01-07.

[25] Zaide, Gregorio F. (1957). *Philippine Political and Cultural History* (http://books.google.com/?id=pJYVAQAAIAAJ&q=Kingdom+of+tondo&dq=Kingdom+of+tondo). Philippine Education Co. p. 42. . Retrieved 2010-01-07.

[26] Zhang Xie. (1618) (in Chinese). *Dong Xi Yang Kao [A Study of the Eastern and Western Oceans] Volume 5* (http://www.lib.kobe-u.ac.jp/directory/sumita/5A-161/volume05.html) (Chinese: (http://www.lib.kobe-u.ac.jp/directory/sumita/5A-161/)). ISBN 7532515931. MID 00024687. Retrieved 2009-12-18.

[27] Bascar, C.M. (n.d.). Sultanate of Sulu, "The Unconquered Kingdom" (http://www.royalsulu.com/history.html). Retrieved 2009-12-19 from The Royal Hashemite Sultanate of Sulu & Sabah Official Website (http://www.royalsulu.com/).

[28] Munoz, Paul Michel (2006). *Early Kingdoms of the Indonesian Archipelago and the Malay Peninsula.* Singapore: Editions Didier Millet. p. 171. ISBN 9814155675.

[29] U.S. Department of State. Bureau of East Asian and Pacific Affairs. (June 2009). Background Note: Brunei (http://www.state.gov/r/pa/ei/bgn/2700.htm). Retrieved 2009-12-18.

[30] Agoncillo, Teodoro A. (1990). *History of the Filipino People* (8th ed.). Garotech Publishing. p. 22. ISBN 9718711066.

[31] McAmis, Robert Day (2002). *Malay Muslims: The History and Challenge of Resurgent Islam in Southeast Asia* (http://books.google.com/?id=59PnSwurWj8C&pg=PA18&dq=#v=onepage&q=). Wm. B. Eerdmans Publishing. pp. 18–24, 53–61. ISBN 0802849458. . Retrieved 2010-01-07.

[32] Zaide, Gregorio F. and Sonia M. Zaide (2004). *Philippine History and Government* (6th ed.). All-Nations Publishing Company.

[33] Kurlansky, Mark. (1999). *The Basque History of the World.* New York: Walker & Company. p. 64. ISBN 0-8027-1349-1.

[34] Joaquin, Nick. (1988). *Culture and History: Occasional Notes on the Process of Philippine Becoming.* Manila: Solar Publishing.

[35] Dolan, Ronald E. (Ed.). (1991). "Education" (http://countrystudies.us/philippines/53.htm). *Philippines: A Country Study* (http://countrystudies.us/philippines/). Washington: GPO for the Library of Congress. Retrieved 2009-12-20 from Country Studies US Website (http://countrystudies.us/).

[36] Halili, Maria Christine N. (2004). *Philippine History* (http://books.google.com/?id=gUt5v8ET4QYC&pg=PA119&q=). Rex Bookstore. pp. 119–120. ISBN 9712339343. . Retrieved 2010-01-08.

[37] De Borja, Marciano R. (2005). *Basques in the Philippines* (http://books.google.com/?id=xXpiujH2uOwC&pg=PA81&q=). University of Nevada Press. pp. 81–83. ISBN 0874175909. . Retrieved 2010-01-08.

[38] Nuguid, Nati. (1972). "The Cavite Mutiny" (http://stuartxchange.com/CaviteMutiny.html). in Mary R. Tagle. *12 Events that Have Influenced Philippine History.* [Manila]: National Media Production Center. Retrieved 2009-12-20 from StuartXchange Website (http://stuartxchange.com/).

[39] Joaquin, Nick. *A Question of Heroes.*

[40] Richardson, Jim. (January 2006). "Andrés Bonifacio Letter to Julio Nakpil, April 24, 1897" (http://kasaysayan-kkk.info/docs.ab.240497.jn.htm). *Documents of the Katipunan* (http://kasaysayan-kkk.info/default.template.0407.htm). . Retrieved 2009-12-19.

[41] Bautista, Veltisezar. (2002). "3. The Philippine Revolution (1896–1898)" (https://bookhaus2.securesites.net/philnewscentral/cgi-bin/redirect.cgi?url=philrev.html). *The Filipino Americans (1763–present): Their History, Culture and Traditions* (https://bookhaus2.securesites.net/philnewscentral/cgi-bin/redirect.cgi?url=filipino_am.html) (2nd ed.). Naperville, IL: Bookhaus Publishers. ISBN 0-931613-17-5. . Retrieved 2009-05-14.

[42] Ocampo, Ambeth. (1999). *Rizal Without the Overcoat* (Expanded ed.). Pasig City: Anvil Publishing, Inc. ISBN 971-27-0920-5.

[43] Price, Michael G. (2002). Foreword. In A. B. Feuer, *America at War: the Philippines, 1898–1913* (http://books.google.com/books?id=f-2Qef1JSjsC&printsec=frontcover&source=gbs_navlinks_s#v=onepage&q=&f=false) (pp. xiii–xvi). Westport, CT: Greenwood. ISBN 0-275-96821-9.

[44] Gates, John M. (November 2002). "The Pacification of the Philippines" (http://www3.wooster.edu/history/jgates/book-ch3.html). *The U.S. Army and Irregular Warfare* (http://www3.wooster.edu/history/jgates/book-contents.html). . Retrieved 2010-02-20.

[45] White, Matthew. "Death Tolls for the Man-made Megadeaths of the 20th Century" (http://users.erols.com/mwhite28/battles.htm#Manila). Retrieved 2007-08-01.

[46] Rottman, Gordon L. (2002). *World War 2 Pacific Island Guide – A Geo-Military Study* (http://books.google.com/books?id=ChyilRml0hcC). Westport, CT: Greenwood Press. p. 318 (http://books.google.com/books?id=ChyilRml0hcC&pg=PA318&dq=#v=onepage&q&f=false). ISBN 0-313-31395-4.

[47] Chandler, David P. and David Joel Steinberg (1987). *In Search of Southeast Asia: A Modern History* (http://books.google.com/?id=jzUz9lKn6PEC&pg=PA431&dq=#v=onepage&q=) (Revised 2nd ed.). University of Hawaii Press. pp. 431–442. ISBN 0824811100. . Retrieved 2010-01-07.

[48] Osborne, Milton E. (2004). *Southeast Asia: An Introductory History* (http://books.google.com/?id=uaFaDUyeCOcC&pg=PA235&dq=#v=onepage&q=) (9th ed.). Allen & Unwin. pp. 235–241. ISBN 1741144485. . Retrieved 2010-01-07.

[49] Tarling, Nicholas (2000). *The Cambridge History of Southeast Asia: From World War II to the Present, Volume 4* (http://books.google.com/?id=U0trzUvic-8C&pg=PA293&dq=#v=onepage&q=). Cambridge University Press. p. 293. ISBN 0521663725. . Retrieved 2010-01-07.

[50] Aquino, Corazon (1996-10-11). *Corazon Aquino Speaks to Fulbrighters* (http://web.archive.org/web/20080626025305/http://www.fulbrightalumni.org/olc/pub/FBA/fulbright_prize/aquino_address.html) (Speech). Washington, D.C.. Archived from the original (http://www.fulbright.org/sites/default/files/Aquino Address.doc) on 2008-06-26. . Retrieved 2010-08-13.

[51] Gargan, Edward A. (December 11, 1997). "Last Laugh for the Philippines; Onetime Joke Economy Avoids Much of Asia's Turmoil" (http://www.nytimes.com/1997/12/11/business/last-laugh-for-philippines-onetime-joke-economy-avoids-much-asia-s-turmoil.html). *New York Times*. . Retrieved 2008-01-25.

[52] Robles, Alan C. (July/August 2008). "Civil service reform: Whose service?" (http://www.inwent.org/ez/articles/077943/index.en.shtml). *D+C* (Internationale Weiterbildung und Entwicklung [InWEnt]) **49**: 285–289. . Retrieved 2008-11-30.

[53] Bigornia, Amante. (1997-09-17). "The 'consultations' on Charter change" (http://news.google.com/newspapers?id=no8VAAAAIBAJ&sjid=bQsEAAAAIBAJ&pg=4208,1807319&dq=). *The Manila Standard*. . Retrieved 2009-12-13.

[54] "Guide to the Philippines conflict" (http://news.bbc.co.uk/2/hi/asia-pacific/1695576.stm). (2007-08-10). *BBC News*. Retrieved 2009-12-16.

[55] World Bank. Conflict Prevention & Reconstruction Unit. (February 2005). *The Mindanao Conflict in the Philippines: Roots, Costs, and Potential Peace Dividend* (http://siteresources.worldbank.org/INTCPR/214578-1111996036679/20482477/WP24_Web.pdf) by Salvatore Schiavo-Campo and Mary Judd. Washington, D.C.: World Bank. (Social Development Paper No. 24). Retrieved 2009-12-16.

[56] Liefer, Michael. (2005). *Michael Liefer – Selected Works on Southeast Asia* (http://books.google.com/books?id=B9kTZgZ0TWYC&printsec=frontcover&source=gbs_navlinks_s#v=onepage&q=&f=false) (Chin, Kin-Wah & Leo Suryadinata, Eds.). Singapore: Institute of Southeast Asian Studies. ISBN 981-230-270-0.

[57] The White House. (2003-03-27). "Coalition Members" (http://georgewbush-whitehouse.archives.gov/news/releases/2003/03/20030327-10.html). . Retrieved 2009-12-18.

[58] U.S. Department of State. Bureau of East Asian and Pacific Affairs. (October 2009). "Background Note: Philippines" (http://www.state.gov/r/pa/ei/bgn/2794.htm). . Retrieved 2009-12-18.

[59] Permanent Mission of the Republic of the Philippines to the United Nations. [ca. 2008]. About Us (http://www.un.int/philippines/about/). Retrieved 2010-08-13.

[60] Permanent Mission of the Republic of the Philippines to the United Nations. [ca. 2008]. The Philippines and the UN Security Council (http://www.un.int/philippines/security_council/). Retrieved 2008-01-12.

[61] United Nations Human Rights Council. [ca. 2009]. Membership by regional groups from 19 June 2006–18 June 2007 (http://www2.ohchr.org/english/bodies/hrcouncil/groups.htm). Retrieved 2010-03-21.

[62] United Nations Security Council. (1999-10-25). *Resolution 1272* (http://daccess-ods.un.org/access.nsf/Get?Open&DS=S/RES/1272 (1999)&Lang=E&Area=UNDOChttp://daccess-ods.un.org/access.nsf/Get?Open&DS=S/RES/1272 (1999)&Lang=E&Area=UNDOC) [**S-RES-1272(1999)**]. Retrieved 2010-03-21.

[63] Bangkok Declaration. (1967-08-08). Retrieved 2009-12-20 from Wikisource.

[64] "ASEAN Primer" (http://web.archive.org/web/20071217132852/http://www.summit99.ops.gov.ph/asean_structure.htm). (1999). *3rd ASEAN Informal Summit*. Archived from the original (http://www.summit99.ops.gov.ph/asean_structure.htm) on 2007-12-17. Retrieved 2009-12-13.

[65] Chua, Yvonne T. and Ellen Tordesillas. (2008-03-09). "6 Philippine-occupied islands covered in Spratly agreements" (http://www.gmanews.tv/story/84023/6-Philippine-occupied-islands-covered-in-Spratly-agreements). *GMA News*. . Retrieved 2009-12-13.

[66] Ministry of Foreign Affairs of Japan. [ca. 2009]. "Japan's ODA Data by Country – Philippines" (http://www.mofa.go.jp/policy/oda/data/pdfs/philippines.pdf). . Retrieved 2010-06-02.

[67] Dolan, Ronald E. (Ed.). (1991). "Relations with Asian Neighbors" (http://countrystudies.us/philippines/93.htm). *Philippines: A Country Study* (http://countrystudies.us/philippines/). Washington: GPO for the Library of Congress. Retrieved 2010-01-05 from Country Studies US Website (http://countrystudies.us/).

[68] "DFA: 'Technicalities' blocking RP bid for OIC observer status" (http://www.gmanews.tv/story/162925/DFA-Technicalities-blocking-RP-bid-for-OIC-observer-status). (2009-05-26). *GMA News*. Retrieved 2009-07-10.

[69] Balana, Cynthia. (May 26, 2009). "RP nears observer status in OIC – DFA" (http://globalnation.inquirer.net/news/news/view/20090526-207265/RP-nears-observer-status-in-OIC----DFA). *The Philippine Daily Inquirer*. . Retrieved 2009-07-10.

[70] Republic of the Philippines. National Statistical Coordination Board. " List of Regions (http://www.nscb.gov.ph/activestats/psgc/listreg.asp)". (March 2010). Retrieved 2010-05-21.

[71] Republic Act No. 5446 – An Act to Amend Section One of Republic Act Numbered Thirty Hundred and Forty-Six, Entitled "An Act to Define the Baselines of the Territorial Sea of the Philippines" (http://www.chanrobles.com/republicacts/republicactno5446.html). Republic of the Philippines. (September 18, 1968). Retrieved 2009-11-21 from the Chan Robles Virtual Law Library (http://www.chanrobles.com/).

[72] Central Intelligence Agency. (2009). "Field Listing :: Coastline" (https://www.cia.gov/library/publications/the-world-factbook/fields/2060.html). Washington, DC: Author. Retrieved 2009-11-07.

[73] Republic of the Philippines. Department of Tourism. [ca. 2008]. Leyte is Famous For... (http://www.travelmart.net/philippines/leyte-hotels-resorts.html). Retrieved 2010-03-21 from www.travelmart.net. Archived from the original (http://www.tourism.gov.ph/

explore_phil/place_details.asp?content=famousefor&province=4).
[74] "Submissions, through the Secretary-General of the United Nations, to the Commission on the Limits of the Continental Shelf, pursuant to article 76, paragraph 8, of the United Nations Convention on the Law of the Sea of 10 December 1982" (http://www.un.org/Depts/los/clcs_new/commission_submissions.htm). United Nations Commission on the Limits of the Continental Shelf. 28 May 2009. . Retrieved 29 May 2009.
[75] La Putt, Juny P. [ca. 2003]. *The 1990 Baguio City Earthquake* (http://www.cityofpines.com/baguioquake/quake.html). Retrieved 2009-12-20 from The City of Baguio (http://www.cityofpines.com/) Website.
[76] Newhall, Chris, James W. Hendley II, and Peter H. Stauffer. (2005-02-28). "The Cataclysmic 1991 Eruption of Mount Pinatubo, Philippines (U.S. Geological Survey Fact Sheet 113-97)" (http://pubs.usgs.gov/fs/1997/fs113-97/). U.S. Department of the Interior. U.S. Geological Survey. . Retrieved 2007-04-09.
[77] Greenlees, Donald. (May 14, 2008). "Miners shun mineral wealth of the Philippines" (http://www.nytimes.com/2008/05/14/business/worldbusiness/14iht-mine.1.12876764.html). *The New York Times*. . Retrieved 2009-12-11.
[78] Davies, Ed and Karen Lema. (2008-06-29). "Pricey oil makes geothermal projects more attractive for Indonesia and the Philippines" (http://www.nytimes.com/2008/06/29/business/worldbusiness/29iht-energy.1.14068397.html). *The New York Times*. . Retrieved 2009-12-18.
[79] "Natural Resources and Environment in the Philippines" (http://www.etravelpilipinas.com/about_philippines/philippine_natural_resources.htm). (n.d.). *eTravel Pilipinas*. Retrieved January 22, 2009.
[80] Chanco, Boo. (1998-12-07). "The Philippines Environment: A Warning" (http://gbgm-umc.org/asia-pacific/philippines/ecophil.html). *The Philippine Star*. . Retrieved 2010-02-15 from gbgm-umc.org.
[81] Williams, Jann, Cassia Read, Tony Norton, Steve Dovers, Mark Burgman, Wendy Proctor, and Heather Anderson. (2001). "The Meaning, Significance and Implications of Biodiversity (continued)" (http://www.environment.gov.au/soe/2001/publications/theme-reports/biodiversity/biodiversity01-3.html). *Biodiversity Theme Report* (http://www.environment.gov.au/soe/2001/publications/theme-reports/biodiversity/index.html). CSIRO on behalf of the Australian Government Department of the Environment and Heritage. ISBN 0643067493. . Retrieved 2009-11-06.
[82] Carpenter, Kent E. and Victor G. Springer. (April 2005). "The center of the center of marine shore fish biodiversity: the Philippine Islands" (http://www.springerlink.com/content/w2k20vqkk553p746/). *Environmental Biology of Fishes* (Springer Netherlands) **74** (2): 467–480. doi:10.1007/s10641-004-3154-4. .
[83] Rowthorn, Chris and Greg Bloom. (2006). *Philippines* (http://books.google.com/?id=aaUR07G0yAcC) (9th ed.). Lonely Planet. p. 52 (http://books.google.com/books?id=aaUR07G0yAcC&pg=PA52). ISBN 1741042895. . Retrieved 2009-12-10.
[84] Regalado, Jacinto C. Jr. and Lawrence R. Heaney. (1998). "Vanishing Treasures" (http://web.archive.org/web/20090318012913/http://www.fieldmuseum.org/vanishing_treasures/V_Mammals.htm). In Lawrence R. Heaney, *Vanishing Treasures of the Philippine Rain Forest*. Chicago: Field Museum. (archived from the original (http://www.fieldmuseum.org/vanishing_treasures/V_Mammals.htm) on 2009-03-18).
[85] BirdLife International. (2004). *Pithecophaga jefferyi*. 2006. *IUCN Red List of Threatened Species*. IUCN 2006. www.iucnredlist.org (http://www.iucnredlist.org). Retrieved on 2009-1-7.
[86] Conservation International. Center for Applied Biodiversity Science. [ca. 2007]. "Philippines" (http://www.biodiversityhotspots.org/xp/Hotspots/philippines/Pages/default.aspx). In *Biodiversity Hotspots*. Retrieved 2009-12-20.
[87] Taguinod, Fioro. (2008-11-20). "Rare flower species found only in northern Philippines" (http://www.gmanews.tv/story/134682/Rare-flower-species-found-only-in-northern-Philippines). *GMA News*. Retrieved 2009-12-14.
[88] de Vera, Ellayn and Charrissa M. Luci. (2005-07-17). "Balete Drive: 'White Lady', 'haunted' houses and other myths". *The Manila Bulletin*.
[89] WWF-Philippines. (2008-09-02). Apo Reef Set to Reclaim Old Glory: park opens for tourism, closes for fishing (http://www.wwf.org.ph/newsfacts.php?pg=det&id=96). Retrieved 2010-04-26 from the WWF-Philippines Website.
[90] "About the Philippines" (http://www.philippine-history.org/about-philippines.htm). (2009-10-17). Retrieved 2009-12-20 from the Philippine History Website.
[91] Peralta, Eleno O. (2005). " 21. Forests for poverty alleviation: the response of academic institutions in the Philippines (http://www.fao.org/docrep/008/af349e/af349e0n.htm#bm23)". In Sim, Appanah, and Hooda (Eds.). *Proceedings of the workshop on forests for poverty reduction: changing role for research, development and training institutions* (RAP Publication). Food and Agriculture Organization (FAO). Retrieved 2009-12-20.
[92] Heaney, Lawrence R. [ca. 2002]. " The Causes and Effects of Deforestation (http://web.archive.org/web/20090317192337/http://www.fieldmuseum.org/vanishing_treasures/Deforestation_1.htm)". *Field Museum of Natural History*. (archived from the original (http://www.fieldmuseum.org/Vanishing_Treasures/Deforestation_1.htm) on 2009-03-17).
[93] Kirby, Alex. (2003-07-23). SE Asia faces 'catastrophic' extinction rate (http://news.bbc.co.uk/2/hi/science/nature/3090071.stm). *BBC News*. Retrieved 2009-12-20.
[94] Philippine Atmospheric, Geophysical and Astronomical Services Administration. (n.d.). "Climate of the Philippines" (http://kidlat.pagasa.dost.gov.ph/cab/cab.htm). . Retrieved 2010-04-24.
[95] Lonely Planet. (n.d.). Philippines: When to go & weather (http://www.lonelyplanet.com/philippines/weather). Retrieved 2009-01-23.
[96] Library of Congress – Federal Research Division. (March 2006). *Country Profile: Philippines* (http://lcweb2.loc.gov/frd/cs/profiles/Philippines.pdf). Retrieved 2009-12-17.
[97] Chong, Kee-Chai, Ian R. Smith, and Maura S. Lizarondo. (1982). "III. The transformation sub-system: cultivation to market size in fishponds" (http://wayback.archive.org/web/*/http://www.unu.edu/unupress/unupbooks/80346e/80346E06.htmhtm). *Economics of*

the Philippine Milkfish Resource System (http://web.archive.org/web/20110719181709/http://archive.unu.edu/unupress/unupbooks/80346e/80346E00.htm). The United Nations University. ISBN 9280803468. Archived from the original (http://www.unu.edu/unupress/unupbooks/80346e/80346E00.htm) on 2011-07-19. . Retrieved 2009-05-14.

[98] Philippine Atmospheric, Geophysical and Astronomical Services Administration (PAGASA). (January 2009). *Member Report to the ESCAP/WMO Typhoon Committee, 41st Session* (http://www.typhooncommittee.org/41st/docs/TC2_MemberReport2008_PHILIPPINES1.pdf). . Retrieved 2009-12-17.

[99] *Monthly Typhoon Tracking Charts* (http://agora.ex.nii.ac.jp/digital-typhoon/reference/monthly/). (2010). Retrieved 2010-04-24 from the National Institute of Informatics, Kitamoto Laboratory, Digital Typhoon Website.

[100] Glossary of Meteorology. Baguio. (http://amsglossary.allenpress.com/glossary/browse?s=b&p=4) Retrieved on 2008-06-11.

[101] Republic of the Philippines. National Statistical Coordination Board. "Third Quarter 2009 Gross National Product and Gross Domestic Product by Industrial Origin" (http://www.nscb.gov.ph/sna/2009/3rdQ2009/2009gnpi3.asp). Author. . Retrieved 2009-12-11.

[102] Republic of the Philippines. National Statistics Office. (October 2009). "Quickstat" (http://www.census.gov.ph/data/quickstat/qs0909tb.pdf) (PDF). Author. . Retrieved 2009-12-11.

[103] Denis Somoso. (2011-11-04). "$75.174 Billion - Philippines GIR now Rank 26th World's highest International Reserves" (http://www.hikot.com/index.php?do=/blog/25/75-174-billion-philippines-gir-now-rank-26th-worldâs-highest-international-/). *Hikot Global Pinoy*. Retrieved 2011-11-05.

[104] The Filipina sisterhood (http://www.economist.com/world/asia/displaystory.cfm?story_id=E1_RRPJDJ). (2001-12-20). *The Economist*. Retrieved 2009-11-09.

[105] Ure, John. (2008). *Telecommunications Development in Asia* (http://books.google.com/?id=rujyOiFMl0MC&printsec=frontcover&q=). Hong Kong University Press. pp. 301–302. ISBN 9789622099036. . Retrieved 2009-11-09.

[106] Felix, Rocel. (2008-01-25). 2007 GDP seen growing at fastest rate in 30 years (http://business.inquirer.net/money/breakingnews/view/20080125-114787/2007-GDP-seen-growing-at-fastest-rate-in-30-years). *The Philippine Daily Inquirer*. Retrieved 2010-05-29.

[107] Philippines Aims to Boost Growth by 2009 (http://web.archive.org/web/19960101-re_/http://www.forbes.com/feeds/ap/2007/02/20/ap3443340.html). (2007-02-20). *Forbes*. Associated Press. Archived from the original (http://www.forbes.com/feeds/ap/2007/02/20/ap3443340.html) on 2007-02-22. Retrieved 2008-01-09.

[108] United Nations Development Programme. (2009). "Table G: Human development and index trends, Table I: Human and income poverty". *Human Development Report 2009 – Overcoming barriers: Human mobility and development* (http://hdr.undp.org/en/media/HDR_2009_EN_Complete.pdf) (Palgrave Macmillan). ISBN 978-0-230-23904-3.

[109] Reddel, Paul. (2009-05-27). *Infrastructure & Public-Private Partnerships in East Asia and the Philippines* (http://www.ppiaf.org/documents/Phil_presentation-May_2009-PR.pdf) [PowerPoint slides]. Presentation in Manila to the American Foreign Chambers of Commerce of the Philippines. Retrieved 2010-02-13 from the Public-Private Infrastructure Advisory Facility (PPIAF) Website.

[110] "Report for Selected Countries and Subjects" (http://www.imf.org/external/pubs/ft/weo/2010/02/weodata/weorept.aspx?pr.x=37&pr.y=3&sy=2005&ey=2010&scsm=1&ssd=1&sort=subject&ds=.&br=1&c=566,536,578,548&s=NGDPD,NGDPDPC&grp=0&a=). Imf.org. 2006-09-14. . Retrieved 2011-10-23.

[111] Beyond 'Imperial Manila' (http://www.newsflash.org/2004/02/ht/ht006433.htm). (2006-07-25). *The Manila Standard Today*. Archived from the original (http://www.manilastandardtoday.com/?page=editorial_july25_2006) on unknown date. Retrieved 2009-12-16.

[112] Llorito, David. (2006-05-10). "Help wanted for Philippines outsourcing" (http://www.atimes.com/atimes/Southeast_Asia/HE10Ae02.html). Asia Times. . Retrieved 2009-12-11.

[113] Wilson, Dominic and Anna Stupnytska. (2007-03-28). "The N-11: More Than an Acronym" (Global Economics Paper No: 153) (http://www.chicagobooth.edu/alumni/clubs/pakistan/docs/next11dream-march '07-goldmansachs.pdf). Goldman Sachs Economic Research. Retrieved 2009-11-07.

[114] Olchondra, Riza T. (2006-10-02). As India gets too costly, BPOs turn to Philippines (http://business.inquirer.net/money/topstories/view_article.php?article_id=24405). *The Philippine Daily Inquirer*. Retrieved 2009-12-16.

[115] CO2 Emissions from Fuel Combustion (http://www.iea.org/co2highlights/co2Highlights.XLS) Population 1971-2008 (pdf (http://iea.org/co2highlights/co2highlights.pdf) pages 83-85) IEA (OECD/ World Bank) (original population ref OECD/ World Bank e.g. in IEA Key World Energy Statistics 2010 page 57)

[116] Republic of the Philippines. National Statistical Coordination Board. Population of the Philippines Census Years 1799 to 2007 (http://www.nscb.gov.ph/secstat/d_popn.asp). Retrieved 2009-12-11.

[117] "Bishops threaten civil disobedience over RH bill" (http://www.gmanews.tv/100days/story/202186/bishops-threaten-civil-disobedience-over-rh-bill). GMA News. 2010-09-29. . Retrieved 2010-10-16.

[118] Central Intelligence Agency. "Field Listing :: Life expectancy at birth" (https://www.cia.gov/library/publications/the-world-factbook/fields/2102.html). Washington, DC: Author. . Retrieved 2009-12-11.

[119] Collymore, Yvette. (June 2003). "Rapid Population Growth, Crowded Cities Present Challenges in the Philippines" (http://www.prb.org/Articles/2003/RapidPopulationGrowthCrowdedCitiesPresentChallengesinthePhilippines.aspx). Population Reference Bureau. . Retrieved 2010-04-26.

[120] Asis, Maruja M.B. (January 2006). " The Philippines' Culture of Migration (http://www.migrationinformation.org/Profiles/display.cfm?ID=364)". *Migration Information Source*. Migration Policy Institute. Retrieved 2009-12-14.

[121] "Selected Population Profile in the United States: Filipino alone or in any combination" (http://factfinder.census.gov/servlet/IPTable?_bm=y&-context=ip&

-reg=ACS_2007_1YR_G00_S0201:038;ACS_2007_1YR_G00_S0201PR:038;ACS_2007_1YR_G00_S0201T:038;ACS_2007_1YR_G00_S0201TPR:038& -qr_name=ACS_2007_1YR_G00_S0201&-qr_name=ACS_2007_1YR_G00_S0201PR&-qr_name=ACS_2007_1YR_G00_S0201T& -qr_name=ACS_2007_1YR_G00_S0201TPR&-ds_name=ACS_2007_1YR_G00_&-tree_id=306&-redoLog=false&-geo_id=01000US& -geo_id=NBSP&-search_results=16000US3651000&-format=&-_lang=en). United States Census Bureau. . Retrieved 2009-02-01. The U.S. Census Bureau 2007 American Community Survey counted 3,053,179 Filipinos; 2,445,126 native and naturalized citizens, 608,053 of whom were not U.S. citizens.

[122] Castles, Stephen and Mark J. Miller. (July 2009). " Migration in the Asia-Pacific Region (http://www.migrationinformation.org/Feature/display.cfm?id=733&feed=rss)". *Migration Information Source*. Migration Policy Institute. Retrieved 2009-12-17.

[123] Ciria-Cruz, Rene P. (2004-07-26). 2 million reasons for withdrawing 51 troops (http://www.sfgate.com/cgi-bin/article.cgi?file=/chronicle/archive/2004/07/26/EDGD56NB0H1.DTL). *San Francisco Chronicle*. July 26, 2004.

[124] Republic of the Philippines. National Statistics Office. (2009). *The Philippines in Figures 2009* (http://www.census.gov.ph/data/publications/pif_2009.pdf). Author. ISSN 1655-2539. . Retrieved 2009-12-23.

[125] " Philippines (http://www.britannica.com/EBchecked/topic/456399/Philippines)". (2009). In *Encyclopædia Britannica*. Retrieved 2009-12-18 from Encyclopædia Britannica Online.

[126] Lewis, Paul M. (2009). Languages of Philippines (http://www.ethnologue.com/show_country.asp?name=PH). *Ethnologue: Languages of the World* (16th ed.). Dallas, Tex.: SIL International. Retrieved 2009-12-16.

[127] Dolan, Ronald E. (Ed.). (1991). "Ethnicity, Regionalism, and Language" (http://countrystudies.us/philippines/35.htm). *Philippines: A Country Study* (http://countrystudies.us/philippines/). Washington: GPO for the Library of Congress. Retrieved 2010-04-08 from Country Studies US Website (http://countrystudies.us/).

[128] Capelli, Christian, James F. Wilson, Martin Richards, Michael P. H. Stumpf, Fiona Gratrix, Stephen Oppenheimer, Peter Underhill et al. (2001-02-01). "A Predominantly Indigenous Paternal Heritage for the Austronesian-Speaking Peoples of Insular South Asia and Oceania" (http://hpgl.stanford.edu/publications/AJHG_2001_v68_p432.pdf). *American Journal of Human Genetics* **68** (2): 432–443. doi:10.1086/318205. PMC 1235276. PMID 11170891. . Retrieved 2009-12-18.

[129] " The Impact of Spanish Rule in the Philippines (http://www.seasite.niu.edu/Tagalog/Tagalog_Homepage99/impacts_of_spanish_rule_in_the_p.htm)". (2009). *Tagalog at NIU*. Retrieved 2009-12-19 from the Northern Illinois University, Center for Southeast Asian Studies, SEAsite Project.

[130] " Chinese lunar new year might become national holiday in Philippines too (http://news.xinhuanet.com/english/2009-08/23/content_11930729.htm)". (2009-08-23). *Xinhua News*. Retrieved 2009-12-18.

[131] Republic of the Philippines. National Statistics Office. (April 2008). "Total Population and Annual Population Growth Rates by Region: Population Censuses 1995, 2000, and 2007" (http://www.census.gov.ph/data/census2007/index.html). . Retrieved 2010-04-04.

[132] Demographia. (July 2010). *Demographia World Urban Areas (World Agglomerations) Population & Projections* (http://www.demographia.com/db-worldua.pdf) (Edition 6.1). Retrieved 2011-03-29.

[133] Republic of the Philippines. National Statistical Coordination Board. (July 2009). *2008 Gross Regional Domestic Product - Levels of GRDP* (http://www.nscb.gov.ph/grdp/2008/2008conlev.asp). Retrieved 2010-04-04.

[134] Hawksworth, John, Thomas Hoehn and Anmol Tiwari. "Global City GDP Rankings 2008-2025" (https://www.ukmediacentre.pwc.com/imagelibrary/downloadMedia.ashx?MediaDetailsID=1562). *UK Economic Outlook November 2009*. PricewaterhouseCoopers. p. 20. . Retrieved 2009-11-20.

[135] Philippine Census, 2000. Table 11. Household Population by Ethnicity, Sex, and Region: 2000.

[136] Republic of the Philippines. National Statistics Office. (2003-02-18). "2000 Census: Additional Three Persons Per Minute" (http://www.census.gov.ph/data/pressrelease/2003/pr0323tx.html). Author. . Retrieved 2008-01-09.

[137] RP closer to becoming observer-state in Organization of Islamic Conference (http://www.philstar.com/Article.aspx?articleId=472375&publicationSubCategoryId=205). (2009-05-29). *The Philippine Star*. Retrieved 2009-07-10, "Eight million Muslim Filipinos, representing 10 percent of the total Philippine population, ...".

[138] U.S. Department of State. (2008). *Philippines: International Religious Freedom Report 2010* (http://www.state.gov/g/drl/rls/irf/2010/148891.htm). Retrieved 2011-05-20, "Islam is the largest minority religion, and Muslims constitute between 5 and 9 percent of the total population."

[139] The Largest Baha'i Communities (http://www.adherents.com/largecom/com_bahai.html). (2005-09-30). Retrieved 2010-04-26 from www.adherents.com.

[140] Republic of the Philippines. Department of Education. (2010-09-23). Fact Sheet – Basic Education Statistics (http://www.deped.gov.ph/cpanel/uploads/issuanceImg/2010 _Sept23.xls). Author. Retrieved 2010-04-17.

[141] Republic of the Philippines. Commission on Higher Education. (August 2010). Information on Higher Education System (http://202.57.63.198/chedwww/index.php/eng/Information). *Official Website of the Commission on Higher Education*. Retrieved 2011-04-17.

[142] Republic of the Philippines. (Approved: August 11, 2001). *Republic Act No. 9155 – Governance of Basic Education Act of 2001* (http://www.chanrobles.com/republicactno9155.html). Retrieved 2009-12-11 from the Chan Robles Virtual Law Library (http://www.chanrobles.com/).

[143] Republic of the Philippines. Department of Education. (n.d.). "Historical Perspective of the Philippine Educational System" (http://www.deped.gov.ph/about_deped/history.asp). Author. Retrieved 2009-12-14.

[144] Esplanada, Jerry E. (2009-07-20). Mainstreaming Madrasah (http://newsinfo.inquirer.net/inquirerheadlines/learning/view/20090720-216304/Mainstreaming-Madrasah). *The Philippine Daily Inquirer*. Retrieved 2010-11-25.

[145] Malipot, Ina H. (2010-11-26). Students stage walkouts against cut in SUC funding (http://www.mb.com.ph/articles/289594/students-stage-walkouts-against-cut-suc-funding). *The Manila Bulletin*. Retrieved 2011-04-16.

[146] Republic of the Philippines. (Approved: 2008-04-29). *Republic Act 9500 – An Act to Strengthen the University of the Philippines as the National University* (http://www.up.edu.ph/pdf2/RA_9500.pdf). Retrieved 2011-04-17 from the University of the Philippines Website.

[147] World Health Organization. (2009). *World Health Statistics 2009* (http://www.who.int/entity/whosis/whostat/EN_WHS09_Full.pdf). Geneva: Author. ISBN 9789241563819. . Retrieved 2009-12-23.

[148] Philippine News Agency. (2009-12-14). "Senate approves proposed 2010 national budget" (http://www.gov.ph/index.php?option=com_content&task=view&id=2002495&Itemid=2). Retrieved 2009-12-18 from the Official Government Portal of the Republic of the Philippines (http://www.gov.ph/index.php).

[149] World Health Organization. (April 2006). Philippines (http://www.who.int/countryfocus/cooperation_strategy/ccsbrief_phl_en.pdf). *Country Cooperation Strategy at a Glance*. Retrieved 2009-12-23.

[150] United States Agency for International Development. (May 2008). *USAID Country Health Statistical Report – Philippines* (http://pdf.usaid.gov/pdf_docs/PNADM408.pdf). Retrieved 2010-04-08.

[151] World Bank. [ca. 2010]. *Transport in the Philippines – Overview* (http://web.worldbank.org/WBSITE/EXTERNAL/COUNTRIES/EASTASIAPACIFICEXT/EXTEAPREGTOPTRANSPORT/0,,contentMDK:20458733~menuPK:2069348~pagePK:34004173~piPK:34003707~theSitePK:574066,00.html). Retrieved 2010-03-18 from the World Bank Website.

[152] Republic of the Philippines. Land Transportation Office. Number of Motor Vehicles Registered (http://www.webcitation.org/5stpn1gqg). (2008-01-29). Author. Retrieved January 22, 2009.

[153] The Philippine Transportation System (http://www.asianinfo.org/asianinfo/philippines/pro-transportation.htm). (2008-08-30). *Asian Info*. Retrieved January 22, 2009.

[154] Strong Republic Nautical Highway (http://www.macapagal.com/gma/initiatives/roro.php). (n.d.). Official Website of President Gloria Macapagal Arroyo. Retrieved January 22, 2009.

[155] Gov't revives Pasig River ferry service (http://www.gmanews.tv/story/30644/Govt-revives-Pasig-River-ferry-service). (2007-02-14). *GMA News*. Retrieved 2009-12-18.

[156] " About PAL (http://www.philippineairlines.com/about_pal/about_pal.jsp)". [ca. 2009]. Philippine Airlines Website (http://www.philippineairlines.com/home/home.jsp). Retrieved 2010-01-09.

[157] State of Hawaii. Department of Transportation. Airports Division. [ca. 2005]. " Philippine Air Lines (http://hawaii.gov/hawaiiaviation/hawaii-commercial-aviation/philippine-air-lines/)". *Hawaii Aviation*. Retrieved 2010-01-09.

[158] Oxford Business Group. (2009). *The Report: Philippines 2009* (http://books.google.com/?id=eY-Oq1IGzdMC&pg=PT98&dq=#v=onepage&q=). Author. p. 97. ISBN 1902339126. . Retrieved 2010-01-09.

[159] "Asia's Fab 50 Companies: PLDT-Philippine Long Distance Telephone" (http://www.forbes.com/lists/2008/37/biz_asiafab08_PLDT-Philippine-Long-Distance-Telephone_YIW4.html). *Forbes*. 3 September 2008. Retrieved 2009-13-14.

[160] Wireless Federation. (2009-03-06). Mobile penetration rate reaches the mark of 75% at 2008-end (Philippines) (http://wirelessfederation.com/news/14569-mobile-penetration-rate-reaches-the-mark-of-75-at-2008-end-philippines/). Retrieved 2010-06-02 from the Wireless Federation Website.

[161] Francisco, Rosemarie. (2008-03-04). Filipinos sent 1 billion text messages daily in 2007 (http://technology.inquirer.net/infotech/infotech/view/20080304-122775/Filipinos-sent-1-billion-text-messages-daily-in-2007/). *The Philippine Daily Inquirer*. Reuters. Retrieved 2009-12-18.

[162] Teves, Oliver. (2007-10-29). Cell phones double as electronic wallets in Philippines (http://www.usatoday.com/tech/wireless/phones/2007-09-29-philippines-cell-phones_N.htm). *USA Today*. Associated Press. Retrieved 2009-12-11.

[163] Special Report: The Global 2000 (http://www.forbes.com/lists/2008/18/biz_2000global08_The-Global-2000_Counrty_10.html). (2008-04-02). *Forbes*. p.10. Retrieved 2009-12-14.

[164] Republic of the Philippines. National Telecommunications Commission. [ca. 2010]. "Broadcast(AM,FM,TV,CATV) – Number of Broadcast and CATV Stations by Region" (http://portal.ntc.gov.ph/wps/portal/!ut/p/.cmd/cs/.ce/7_0_A/.s/7_0_C7R/_s.7_0_A/7_0_C7R). Author. . Retrieved 2010-01-16.

[165] Republic of the Philippines. National Telecommunications Commission. [ca. 2010]. "Internet Service Providers – Internet Service" (http://portal.ntc.gov.ph/wps/portal/!ut/p/.cmd/cs/.ce/7_0_A/.s/7_0_EBH/_s.7_0_A/7_0_EBH). Author. . Retrieved 2010-01-16.

[166] Internet World Stats. (2009). Philippines: Internet Usage Stats and Marketing Report (http://www.internetworldstats.com/asia/ph.htm). Miniwatts Marketing Group. Retrieved January 22, 2009.

[167] Universal McCann. *Power To The People: Social Media Tracker, Wave3* (http://www.archive.org/details/UniversalMccannWave3PowerToThePeople). (March 2008). Author. Archived from the original (http://www.universalmccann.com/Assets/wave_3_20080403093750.pdf). Retrieved 2010-03-21.

[168] Liao, Jerry. (2008-05-09). "The Philippines – Social Networking Capital of the World" (http://asia.cnet.com/blogs/infochat/post.htm?id=63003580). Cnet Asia. . Retrieved 2009-11-08.

[169] Baringer, Sally E. [ca. 2006]. "The Philippines" (http://www.everyculture.com/No-Sa/The-Philippines.html). In *Countries and Their Cultures*. Advameg Inc. Retrieved 2009-12-20 from www.everyculture.com.

[170] Rowthorn, Chris and Greg Bloom. (2006). *Philippines* (http://books.google.com/?id=aaUR07G0yAcC&pg=PA44&q=) (9th ed.). Lonely Planet. p. 44. ISBN 1741042895. . Retrieved 2009-12-10.

[171] Dumont, Jean-Paul. (1992). *Visayan Vignettes: Ethnographic Traces of a Philippine Island* (http://books.google.com/?id=zMFKs8--FDMC&pg=PA160&dq=#v=onepage&q=). Chicago: University of Chicago Press. pp. 160–162. ISBN 0226169545. . Retrieved 2010-01-06.

[172] United Nations Educational, Scientific and Cultural Organization. (2010). "Baroque Churches of the Philippines" (http://whc.unesco.org/en/list/677/multiple=1&unique_number=801). *UNESCO World Heritage Centre*. . Retrieved 2010-01-12.

[173] Rowthorn, Chris and Greg Bloom. (2006). *Philippines* (http://books.google.com/?id=aaUR07G0yAcC&pg=PA145&q=) (9th ed.). Lonely Planet. p. 145. ISBN 1741042895. . Retrieved 2010-01-09.

[174] "The Jollibee Phenomenon" (http://web.archive.org/web/20070623034806/http://www.jollibee.com.ph/corporate/phenomenon.htm). Jollibee Inc. Archived from the original (http://www.jollibee.com.ph/corporate/phenomenon.htm) on 2007-06-23. . Retrieved 2008-01-09.

[175] Conde, Carlos H. (2005-05-31). "Jollibee stings McDonald's in Philippines" (http://www.nytimes.com/2005/05/30/business/worldbusiness/30iht-burger.html?pagewanted=all). *The New York Times*. . Retrieved 2010-01-05.

[176] Zialcita, Fernando Nakpil. (2005). *Authentic Though not Exotic: Essays on Filipino Identity* (http://books.google.com/?id=PEy9G-HLokwC&pg=PA281&dq=#v=onepage&q=). Quezon City: Ateneo de Manila University Press. p. 281. ISBN 9715504795. . Retrieved 2010-01-05.

[177] Rowthorn, Chris and Greg Bloom. (2006). *Philippines* (http://books.google.com/?id=aaUR07G0yAcC&pg=PA48&dq=filipino+cuisine&q=filipino cuisine) (9th ed.). Lonely Planet. p. 48. ISBN 1741042895. . Retrieved 2009-12-10.

[178] Zibart, Eve. (2001). *The Ethnic Food Lover's Companion: Understanding the Cuisines of the World* (http://books.google.com/?id=y6vTun3i4NQC&pg=PA266&q=). Menasha Ridge Press. p. 277. ISBN 0897323726. . Retrieved 2009-12-14.

[179] Lopez, Mellie Leandicho. (2006). *A Handbook of Philippine Folklore* (http://books.google.com/?id=jGssp-oJrT8C&printsec=frontcover&q=). University of the Philippines Press. ISBN 9715425143. . Retrieved 2010-01-12.

[180] Republic of the Philippines. National Commission for Culture and the Arts. The National Artists of the Philippines (http://www.ncca.gov.ph/about-ncca/org-awards/org-awards-national-artist-list.php). Retrieved 2009-12-26 from the National Commission for Culture and the Arts Website.

[181] Country profile: The Philippines (http://news.bbc.co.uk/2/hi/asia-pacific/country_profiles/1262783.stm). (2009-12-08). *BBC News*. Retrieved 2009-12-20.

[182] Santiago, Erwin. (2010-04-12). AGB Mega Manila TV Ratings (April 7–11): *Agua Bendita* pulls away (http://www.pep.ph/news/25288/AGB-Mega-Manila-TV-Ratings-(April-7-11):-Agua-Bendita-pulls-away). Retrieved 2010-05-23 from the Philippine Entertainment Portal Website.

[183] "The Philippines' celebrity-obsessed elections" (http://www.economist.com/world/asia/displaystory.cfm?story_id=9084791). (2007-04-26). *The Economist*. Retrieved 2010-01-15.

[184] Viera, Odete Maria and Swarnalata Vemuri. [ca. 2002]. Philippines: Sports and Recreation (https://www.cia.gov/library/publications/world-leaders-1/world-leaders-p/philippines.html). *Cultural Profiles Project*. Citizenship and Immigration Canada (CIC). Retrieved 2009-12-20.

[185] Mga Larong Kinagisnan (http://web.archive.org/web/20071106053308/http://www.hagonoy.com/lbcorpus/halbc010.html) [Games One Grows Up With]. Hagonoy.com. (archived from the original (http://www.hagonoy.com/lbcorpus/halbc010.html) on 2007-11-06)

[186] Mga Larong Pilipino (http://www.seasite.niu.edu./Tagalog/Filipino_Games/mga_larong_pilipino.htm) [Philippine Games]. (2009). *Tagalog at NIU*. Retrieved 2009-12-19 from the Northern Illinois University, Center for Southeast Asian Studies, SEAsite Project.

[187] Yo-yo (http://www.merriam-webster.com/dictionary/yo-yo). (2010). In Merriam-Webster Online Dictionary. Retrieved 2010-01-10.

[188] Republic of the Philippines. (Approved: 2009-12-11). An Act Declaring Arnis as the National Martial Art and Sport of the Philippines (http://www.senate.gov.ph/republic_acts/ra 9850.pdf).Retrieved 2010-02-18 from the Senate of the Philippines Website.

[189] "Billiard Congress of America: Hall of Fame Inductees" (http://home.bca-pool.com/displaycommon.cfm?an=1&subarticlenbr=30). (2009). Retrieved 2009-12-20 from the Billiard Congress of America Website.

[190] "9 named to Philippine Sports Hall of Fame" (http://www.mb.com.ph/articles/254403/9-named-philippine-sports-hall-fame). (2010-04-24). *The Manila Bulletin*. Retrieved 2010-04-24.

[191] Mga Kilalang Pilipino (http://www.seasite.niu.edu/Tagalog/modules_in_Tagalog/mga_kilalang_pilipino.htm) [Known Filipinos]. (n.d.) (in Filipino). *Tagalog at NIU*. Retrieved 2010-04-25 from the Northern Illinois University, Center for Southeast Asian Studies, SEAsite Project.

External links

Government

- Official website of the Official Gazette of the Philippines (http://www.gov.ph/)
- Official website of the House of Representatives of the Philippines (http://www.congress.gov.ph/)
- Official website of the Supreme Court of the Philippines (http://www.judiciary.gov.ph/)
- Official website of Bangko Sentral ng Pilipinas (Central Bank of the Philippines) (http://www.bsp.gov.ph/)
- Official website of the National Economic and Development Authority (NEDA) (http://www.neda.gov.ph/)
- Official website of the Bureau of Agricultural Statistics (http://www.bas.gov.ph/)
- Official website of the Philippine National Police (PNP) (http://www.pnp.gov.ph/)
- Official website of the Department of Tourism (http://www.tourism.gov.ph/)

General information

- Chiefs of State and Cabinet Members (https://www.cia.gov/library/publications/world-leaders-1/world-leaders-p/philippines.html)
- Philippines (http://www.ucblibraries.colorado.edu/govpubs/for/philippines.htm) at UCB Libraries GovPubs
- WikiAnswers: Q&A about the Philippines (http://www.wiki.answers.com/Q/FAQ/2802)
- Philippines (http://www.dmoz.org/Regional/Oceania/Philippines/) at the Open Directory Project
- Philippine News and Current Events (http://www.noypi.ph/)

Books and articles

- History of the Philippine Islands (http://www.gutenberg.org/browse/authors/b#a2296) in many volumes, from Project Gutenberg (indexed under Emma Helen Blair, the general editor)
- Washington Post – How the Philippines Sees America (http://newsweek.washingtonpost.com/postglobal/america/philippines/)

Other

- Asian Development Bank (ADB) (http://www.adb.org/)
- AER – Economic and social issues in the Philippines (http://www.aer.ph/) at Action For Economic Reforms
- Filipinana.net – Free digital library and a research portal (http://www.filipiniana.net/)

Wikimedia

- Wikimedia Philippines
- Wikimedia Atlas of Philippines
- WikiSatellite view of Philippines (http://www.wikimapia.org/#y=12554564&x=122915039&z=6&l=0&m=a) at WikiMapia
- Philippines travel guide from Wikitravel

bjn:Pilipina kbd:Филипинхэр

Palau

Republic of Palau *Beluu ęr a Belau*		

Anthem: *Belau loba klisiich er a kelulul*		
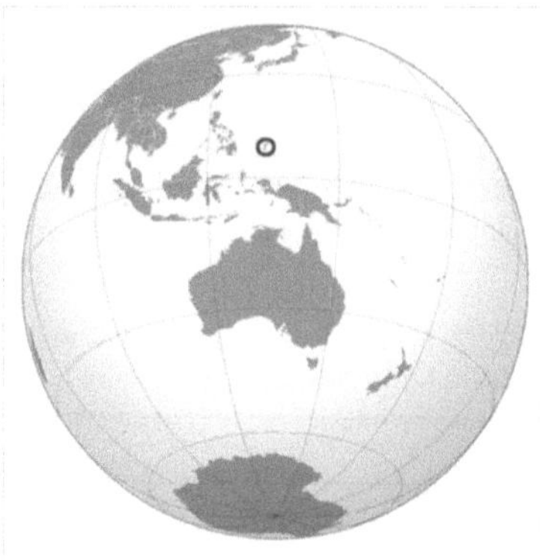 Palau is marked in **green** and in turn **circled** in **green** for better identification.		
Capital		Melekeok[1] 7°21′N 134°28′E
Largest city		Koror
Official language(s)		English, Palauan
Recognised regional languages		Japanese Sonsorolese (in Sonsoral) Tobian (in Hatohobei)
Demonym		Palauan
Government		Unitary presidential Democratic republic
-	President	Johnson Toribiong
-	Vice President	Kerai Mariur
Independence from UN Trust Territory status		
-	Compact of Free Association	October 1, 1994
Area		
-	Total	459 km^2 (196th) 177 sq mi
-	Water (%)	negligible
Population		

-	2011 estimate	20,956 (218th)
-	Density	28.4/km^2 45.5/sq mi
GDP (PPP)		2008 estimate
-	Total	$164 million (2008 est.)[2] (not ranked)
-	Per capita	$8,100[2] (119)
HDI (2011)		0.782[3] (high) (49th)
Currency		US dollar (USD)
Time zone		(UTC+9)
Drives on the		right
ISO 3166 code		PW
Internet TLD		.pw
Calling code		+680
1	On 7 October 2006, government officials moved their offices in the former capital of Koror to Ngerulmud in State of Melekeok, located 20 km (12 mi) northeast of Koror on Babelthaup Island and 2 km (1 mi) northwest of Melekeok village.	
2	GDP estimate includes US subsidy (2004 estimate).	

Palau English pronunciation: /pəˈlaʊ/ (listen), officially the **Republic of Palau** (Palauan: *Beluu ęr a Belau*), is an island nation in the Pacific Ocean, 500 miles (800 km) east of the Philippines and **unknown operator: u','** miles (**unknown operator: u'strong'unknown operator: u',**'km) south of Tokyo. In 1978, after three decades as being part of the United Nations trusteeship, Palau chose independence instead of becoming part of the Federated States of Micronesia, a Compact of Free Association was approved in 1986 but not ratified until 1993. It was put into force the following year, making it one of the world's youngest and smallest sovereign states. In English, the name is sometimes spelled **Belau** in accordance with the native pronunciation. It was formerly also spelled **Pelew**.[4]

History

The archipelago is also known as "The Black Islands." Vintage maps and village drawings, as well as photos of the tattooed and pierced Ibedul of Koror and Lundee, can be found at the Australian Library Online listed in the external links section of this article.

Palau was initially settled over 3,000 years ago, and perhaps 4,500 years ago, probably by migrants from the Philippines. A pygmy population is attested until about 900 years ago. The modern population, judging by its language, may have come from the Sunda Islands. British traders became prominent visitors in the 18th century, followed by expanding Spanish influence in the 19th century. Following its defeat in the Spanish-American War, Spain sold Palau and most of the rest of the Caroline Islands to Germany in 1899. Control passed to Japan in 1914 and during World War II the islands were taken by the United States in 1944, with the costly Battle of Peleliu between September 15 and November 25 when more than 2,000 Americans and 10,000 Japanese were killed. The islands passed formally to the United States under United Nations auspices in 1947 as part of the Trust Territory of the Pacific Islands.

Four of the Trust Territory districts formed a single federated Micronesian state in 1979, but the districts of Palau and the Marshall Islands declined to participate. Palau, the westernmost cluster of the Caroline Islands, instead opted for independent status in 1978, approved a new constitution and became the Republic of Palau in 1981, and signed a Compact of Free Association with the United States in 1982. After eight referendums and an amendment to the

Palauan constitution, the Compact was ratified in 1993 and went into effect on October 1, 1994, marking Palau independent de jure (after Palau was independent de facto since May 25, 1994, when the trusteeship was cancelled).

Legislation making Palau an "offshore" financial center was passed by the Senate in 1998. In 2001, Palau passed its first bank regulation and anti-money laundering laws.

Politics and government

Palau's politics takes place in a multi-party framework of a presidential representative democratic republic, whereby the President of Palau is both head of state and head of government. Executive power is exercised by the government, while legislative power is vested in both the government and the Palau National Congress. The judiciary is independent of the executive and the legislature.

Palau adopted its own constitution in 1981, and the governments of the United States and Palau concluded a Compact of Free Association in 1986, similar to compacts that the United States had entered into with the Federated States of Micronesia and the Republic of the Marshall Islands.[5] The compact entered into force on October 1, 1994, concluding Palau's transition from trusteeship to independence[5] as the last portion of the Trust Territory of the Pacific Islands to gain its independence.

Free association

The Compact of Free Association between the United States and Palau[6] sets forth the free and voluntary association of their Governments, and is primarily focused on the issues of government relations, economic relations, and security and defense relations.[7] Palau has no independent military, and relies on the United States for its defense. Under the Compact, the American military has been granted access to the islands for 50 years. The role of the US Navy is quite minimal, limited to a handful of Navy Seabees (construction engineers) but the United States Coast Guard does have a stronger presence in patrolling the waters.

Foreign relations

As a sovereign nation, Palau conducts its own foreign relations.[5] Since independence, Palau has established diplomatic relations with a number of nations, including many of its Pacific neighbors. Palau was admitted to the United Nations on December 15, 1994, and has since joined several other international organizations. In September 2006, Palau hosted the first Taiwan-Pacific Allies Summit, and its President has gone on several official visits to other Pacific countries, including the Republic of China (Taiwan).

The United States maintains the usual diplomatic delegation and an embassy in Palau, but most aspects of the two countries' relationship have to do with Compact-funded projects, which are the responsibility of the U.S. Department of the Interior's Office of Insular Affairs.[8] This has led to some ambiguity in the official status of Palau, though regarded as de jure independent.

Since 2004, Palau has joined the United States and Israel as the only nations voting consistently against an annual U.N. resolution condemning the United States embargo against Cuba which has been in place since 1962.

On October 5, 2009 Palau formalized Diplomatic and Trade Ties with Malaysia and Lord Morris Davidson was appointed as Palau's first Honorary Consul to Malaysia.

Palau is a member of the Parties to the Nauru Agreement.[9]

Nuclear-free constitution

In 1981, Palau voted for the world's first nuclear-free constitution. This constitution banned the use, storage, and disposal of nuclear, toxic chemical, gas, and biological weapons without first being approved by a 3/4 majority in a referendum.[10] This ban held up Palau's transition to independence because while negotiating a Compact of Free Association with the United States. The U.S. insisted on the option to operate nuclear propelled vessels and store nuclear weapons within the territory.[11] After several referendums that failed to achieve a 3/4 majority, the people of Palau finally approved the compact with the U.S. in 1994.[12] [13]

Asylum for former Guantanamo prisoners

On June 10, 2009, Palau announced that it would accept up to all 17 of the remaining Uyghurs detained in Guantanamo "as a humanitarian gesture."[14] [15] [16] [17] [18] [19] Five Uyghur captives were released without being prosecuted in 2004, and transferred to Albania in 2006. The remaining seventeen were released due to lack of evidence in 2008.

Stuart Beck, Palau's representative to the United Nations, published an op-ed in the *New York Times* disputing the widely distributed report that Palau's agreement to accept the Uyghurs was tied to Palau's receipt of US$200 million in foreign aid.[14] [15] [16] [17] [18] [19] [20] Beck asserted that Palau was receiving funds from the USA connected with accepting the Uyghurs, but it was $90,000 per Uyghur, not "$12 million". Beck asserted that the $90,000 was to compensate Palau for its "relocation costs" and was *"to cover transportation, food, housing and medical help until the men can get oriented and get jobs"*. *ABC News* confirmed that the United States would be paying Palau for each captive who was resettled and quoted a senior official who described the amount as "not very substantial".[21]

William Cleary, writing in the *Pacific Daily News*, reported that Palau was in a poor bargaining position to decline the American request.[22] He wrote:

> "The detainee transfer deal comes at a time when generous U.S. financial aid and economic development grants to Palau under a 15 year old treaty of free association between the two nations have expired. Palauan leaders apparently wanted to demonstrate their commitment to sharing mutual interests and burdens with the USA, as they make the case for renewal of U.S. grant assistance."

However, Cleary also described the assertions that the captive resettlement was a straight *quid pro quo* as "misleading".[22] He listed the USA's desire to continue to be able to base DoD personnel on Palau, to balance the growing military force of the People's Republic of China as one complicating factor. He also listed Palau's desire to have exemption from growing pressure within the USA to restrict immigration from former protectorates like Palau as another complicating factor.

An official "technical working group" from Palau traveled to Guantanamo to meet with the Uyghurs in mid-June 2009.[23] [24] On June 23, the Palau government published a press release which said only eight of the thirteen Uyghurs agreed to meet with the delegation. Their report stated: *"The team advised President Toribiong that in their opinion only a few of the detainees that were interviewed had any real interest in being relocated to Palau."* On June 30, *Radio New Zealand International* reported only one Uyghur agreed to be temporarily resettled in Palau.[25] On September 10, *The Times* reported that three of the Uyghurs, have accepted the invitation to be transferred to asylum in Palau.[26] On September 19, *Fox News* reported that in the week since the first announcement three further Uyghurs agreed to be transferred to Palau.[18] Fox reported that five of the other Uyghurs had refused to speak with Palau officials. On October 31, six Uyghurs were reported to have been transferred to Palau.[27] [28] [29] [30] [31] Twelve of the thirteen remaining Uyghurs were offered asylum. The thirteenth man was not offered asylum because his mental health had deteriorated too severely for the mental health resources available in Palau.

The USA agreed to give Palau additional aid in January 2010.[32] Palau had rejected an earlier aid package of $156 million. The new aid package was for $250 million. President Toribiong asserted that the increase in aid was unrelated to Palau agreement to host the Uyghurs.

States

Palau is divided into sixteen states (until 1984 called municipalities). These are listed below with their areas (in square kilometres) and 2005 Census populations:

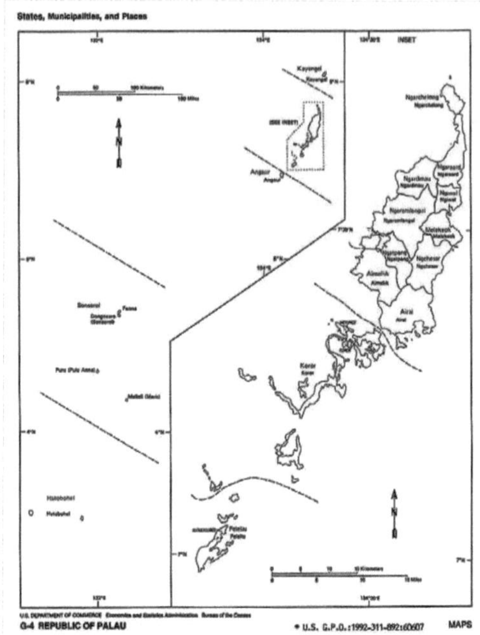

The sixteen states of Palau.

State	Area (km^2)	Census 2005
Aimeliik	52	270
Airai	44	2,723
Angaur	8	320
Hatohobei	3	44
Kayangel	3	188
Koror	18	12,676
Melekeok	28	391
Ngaraard	36	581
Ngarchelong	10	488
Ngardmau	47	166
Ngaremlengui	65	317
Ngatpang	47	464
Ngchesar	41	254
Ngiwal	26	223
Peleliu	13	702
Sonsorol	4	100

Historically, Palau's uninhabited Rock Islands have been part of the State of Koror.

Geography

Palau's most populous islands are Angaur, Babeldaob, Koror, and Peleliu. The latter three lie together within the same barrier reef, while Angaur is an oceanic island several miles to the south. About two-thirds of the population live on Koror. The coral atoll of Kayangel is situated north of these islands, while the uninhabited Rock Islands (about 200) are situated to the west of the main island group. A remote group of six islands, known as the Southwest Islands, some 375 miles (604 km) from the main islands, are also part of the country and make up the states of Hatohobei and Sonsorol.

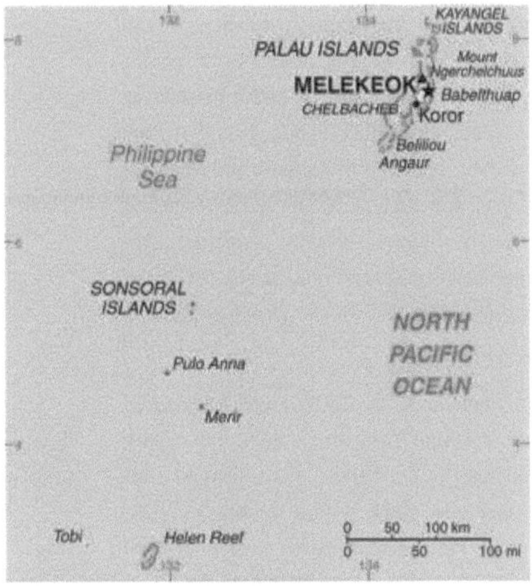

Republic of Palau.

Palau has a tropical climate all year round with an annual mean temperature of 82 °F (28 °C). Rainfall is heavy throughout the year, averaging a total of 150 inches (3,800 mm). The average humidity over the course of the year is 82%, and although rain falls more frequently between July and October, there is still much sunshine. Typhoons are rare, as Palau is outside the main typhoon zone.

Environment

Rock Islands in Palau.

While much of Palau's natural environment remains free of environmental degradation, there are several areas of concern, including illegal fishing with the use of dynamite, inadequate facilities for disposal of solid waste in Koror, and extensive sand and coral dredging in the Palau lagoon. Like the other Pacific island nations, a potential major environmental threat is rising sea levels. Water coverage of low-lying areas is a threat to coastal vegetation, agriculture, and the purity of the nation's water supply. Palau also has a problem with inadequate water supply and limited agricultural areas to support the size of the population. The nation is also vulnerable to earthquakes, volcanic activity, and tropical storms. Sewage treatment is a problem, along with the handling of toxic waste from fertilizers and biocides.

On November 5, 2005, President of Palau, Tommy E. Remengesau, Jr. took the lead on a regional environmental initiative called the Micronesia challenge, which would conserve 30% of near shore coastal waters and 20% of forest land by 2020. In addition to Palau, the initiative was joined by the Federated States of Micronesia and Marshall Islands, and the U.S. territories of Guam and Northern Mariana Islands. Together, this combined region represents nearly 5% of the marine area of the Pacific Ocean and 7% of its coastlines.

WCTC Shopping Center

On September 25, 2009, Palau announced that it would create the world's first "shark sanctuary".[33] Palau has banned all commercial shark fishing within its EEZ waters. The sanctuary protects about 600000 square kilometres (sq mi) of ocean,[34] a similar size to the European country of France.[35] [36] [37] President Johnson Toribiong made the announcement at a meeting of the United Nations.[35] [38] [39] President Toribiong also requested a worldwide ban on fishing for sharks.[35]

Aerial view of limestone islands

Saltwater crocodiles are also residents of Palau and occur in varying numbers throughout the various mangroves and even in parts of the beautiful rock islands. Although this species is generally considered extremely dangerous, there has only been one fatal human attack in Palau within modern history, and that was in the 1960s. In Palau the largest crocodile measured in at 4.5 metres (15 ft).

Economy

Koror-Babeldaob Bridge.

Palau's economy consists primarily of tourism, subsistence agriculture, and fishing. Tourist activity focuses on scuba diving and snorkeling in the islands' rich marine environment, including its barrier reefs walls and World War II wrecks. The government is the major employer of the work force, relying heavily on financial assistance from the US. Business and tourist arrivals numbered 50,000 in the financial year 2000/2001. The population enjoys a per capita income twice that of Micronesia. Long-term prospects for the key tourist sector have been greatly bolstered by the expansion of air travel in the Pacific, the rising prosperity of leading East Asian countries, and the willingness of foreigners to finance infrastructure development.

In July 2004, Palau Micronesia Air was launched with service from Palau to Yap, Guam, Micronesia, Saipan, Australia, and the Philippines. By offering low fares it was planned to be a competitor of Continental Micronesia, however it ceased operations in December of the same year, mainly because of rising fuel prices. Palau Micronesia Air has not restarted operations since but has made a codeshare alliance with Asian Spirit, a carrier that operated flights between Palau and the Philippines (Davao, Cebu, and Manila). There were two flights weekly from Manila via Cebu to Palau and one flight weekly from Davao. Just after few months of service, Asian Spirit ceased its Philippines to Palau route. In March 2010, Pacific Flier began providing air services to Clark, Gold Coast and Guam, however, Pacific Flier suspended operations shortly after, in August of 2010.[40]

In November 2006, the Palau Saving Bank officially announced bankruptcy. On December 13 of the same year the Palau Horizon reported that there were a total of 641 depositors that had been affected by the event. Among the 641 accounts, 398 held less than $5000 USD, with the remainder ranging from $5000 to 2 million USD. On December 12, 79 of the affected people received compensation, one was Taiwanese, while the rest were Palauan, Filipino and US citizens. Mr. Toribiong said, "The fund for the payout came from the balance of Palau government's loan from Taiwan." From a total of $1 million USD, which originally was for assisting Palau's development, there was $955,000 USD left at the time of bankruptcy. For the $955 thousand, Toribiong requested the Taiwanese government use it to repay its loans. Taiwan agreed to the request. The compensation will start its paying to depositor, who held less than $4,000 USD in the account.[41]

Taxes are moderate, the income tax has 3 brackets with medium rates (9.3%, 15% and 19.6%), corporate tax is 4% and general sales tax is 7.5%. There are no property taxes.

Demographics

A deserted island in Palau.

Historical populations	
Year	**Pop.**
1970	11210
1980	12116
1990	15122
2000	21000

The population of Palau is approximately 21,000, of whom 70% are native Palauans, who are of mixed Melanesian, Micronesian, and Austronesian descent. Many Palauans also have some Asian ancestry, which is the result of intermarriage between settlers and Palauans between the 19th and 20th centuries. Palauans with mixed Japanese ancestry accounted for the largest group, and some also had some Chinese or Korean ancestry. Filipinos form the second largest ethnic group.

The official languages of Palau are Palauan and English, except for two states (Sonsorol and Hatohobei) where the local language, along with Palauan, is official. Japanese is also spoken widely amongst older Palauans, and is an official language in the State of Angaur.[42] [43] Tagalog is not official in Palau, but it is the fourth largest spoken language.

Religion

The German and Japanese occupations of Palau both subsidized missionaries. Three quarters of the population are Christians (mainly Roman Catholics and Protestants), while Modekngei (a combination of Christianity, traditional Palauan religion and fortune telling) and the ancient Palauan religion are commonly observed. According to the 2005 census[44] 49.4% of the population is Roman Catholic, 21.3% Protestant, 8.7% Modekngei and 5.3% Seventh-day Adventist. There is a small Jewish community in Palau. In 2009 it sent 3 members to the 18th Maccabiah Games. There are also approximately 400 Bengali Muslims in Palau,[45] and recently 6 Uyghurs detained in Guantanamo Bay were allowed to settle in the island nation.[46]

Culture

Society

Palauan society, much like the island's language, has always been one unique to the island and its people. A very noticeable aspect of Palauan society is that it follows a very strict matrilineal system. Matrilineal practices are seen in nearly every aspect of Palauan traditions, especially in funeral, marriage, inheritance, and the passing of traditional titles.

To this day, the Palauan people still hold true to their traditions very seriously. This is very clear in the fact that the traditional government still holds extreme influence over the nation's affairs. In fact, the traditional government has held so much influence, that the federal government has had, on numerous occasions, attempts at limiting its power. Many of these attempts occurred and continue to occur, from 1990 to the present. These attempts, many of which in the form of amendments in the constitution, were put into place because of the corporate sector of the nation, they having felt that the traditional government was encroaching on what they deemed should be free economic zones. One such example occurred in early 2010, where the Idid clan, the ruling clan of the Southern Federation, under the leadership of Bilung, the clan's and Palau's Southern Federation's queen, raised a civil suit against the KSPLA (Koror State Public Lands Authority). In the civil suit, the Idid clan laid claim over Malakal Island, a major economic zone and Palau's most important port, citing claims that went back as far as the German Era. The civil suit, however, ended with the verdict that Idid clan could not use such citations and claims, and resulted in the conclusion that Malakal Island was land that belonged to the KSPLA.

Traditional Government

The present day traditional government of Palau is a direct continuation of the ancient traditional government, composed of practices that span thousands of years. In the traditional government, Palau is divided geographically into different categories. At the smallest level of geographic division is the village or hamlet, then the chiefdom (which is now politically referred to as a state), and finally the federation, or alliance of chiefdoms. In ancient times, there were numerous federations, or alliances, but upon the introduction of firearms by the British in the 17th century, a major of imbalance of power occurred. Palau was divided into just two major federations, the northern and southern federations. The Northern Federation is headed by the high chief and chiefess of the ruling clan Uudes of Melekeok state, the Reklai and Ebilreklai. As a result of their position, they are commonly referred to as the king and queen of the Northern Federation. This northern federation comprises the following states: Kayangel, Ngerchelong, Ngardmau, Ngiwal, Ngaraard, Ngatpang, Ngeremlengui, Melekok, Aimeliik, Ngchesar, and Airai. The Southern Federation is likewise represented by the high chief and chiefess of the ruling Idid of Koror state, which also results in their titles as king and queen of the Southern Federation. The Southern Federation comprises the following states: Koror, Peleliu, and Angaur. Despite the presence of these terms however, lesser and lesser Palauans have knowledge of the concept of federations, and the term is slowly dying out. Federations had been established as a way of safeguarding states and hamlets who shared economic, social, and political interests, but now with the advent of modernism and a federal government, there is no need for such safeguarding. It is interesting to note however, that in international relations, the king of Palau is often synonymous with the Ibedul of Koror. This is a result of the fact that Koror is the industrial capital of the nation, and because of such, his position and reputation among the corporate sector of the country has a much greater impact than that of the Reklai of Melekeok.

There is also a misconception that the king and queen of Palau, or any chief and his female counterpart for that matter, are married. This is not the case in Palauan society. Traditional leaders and their female counterparts, have always been related and unmaried (marrying relatives in Palauan society has always been a traditional taboo). Usually, a chief and his female counterpart are either brother and sister, or are close cousins, and have their own spouses.

Sports

Baseball is a popular sport in Palau and was introduced to the islands by the Japanese in the 1920s. The Palau national baseball team won the gold medal at the 1990, 1998 and 2010 Micronesian Games, as well as at the 2007 Pacific Games.

Palau also hosts the Palau national football team which competes in the Oceania Football Confederation.

Education

Some fields of study are available at Palau Community College. For professional and graduate programs, students must travel to a larger institution.

Libraries and museums

There are several libraries in Koror, including a public library with a collection comprising about 17,000 books. The Belau National Museum, established in 1956, is also located in Koror and has an affiliated Research Library. Palau Community College also houses a library. In addition to the National Museum, the Etpison family has also opened the Etpison Museum in Koror, which contains many culturally important artifacts.

Transport

Palau International Airport provides scheduled direct flights to Guam, Manila, Seoul and Taipei and Delta Air Lines launched new direct flights to Tokyo-Narita in December 2010. In addition, the states of Angaur and Peleliu have regularly served international airports. Freight, military and holiday cruise ships often call at Malakal Harbor, on Malakal Island outside Koror. The country has no railways, and almost half of the roads are unpaved (of the 61 km/38 mi of highways, only 36 km/22 mi are paved). Driving is on the right and the speed limit is 40 km/h (25 mph). Taxis are available in Koror. They are not metered and fares are negotiable. Only Koror maintains a bus service. Transportation between islands mostly relies on private boats and domestic air services.

Media

- The 1968 World War II movie *Hell in the Pacific* starring Lee Marvin and Toshirō Mifune was filmed on Palau's Rock Islands.
- Palau was the site for the reality television program *Survivor.* Japanese *Survivor* used the islands for its first season in 2002. It was also the location of *Survivor: Palau,* the tenth edition of the American *Survivor,* which aired from February 17 to May 15, 2005 in United States, and *Survivor: Micronesia,* the sixteenth edition, which aired from February 7 to May 11, 2008. In August 2009, *Survivor Philippines* also used the islands for its second season.

Logo for the tenth season of "Survivor." The U.S. media exposure helped to boost tourism in Palau.

- The islands are also mentioned in the song "Orinoco Flow" performed by the Irish singer Enya.
- Palau was featured in an episode of the children's TV show *Toot and Puddle,* which was based on the book series by American children's author Holly Hobbie.
- Palau has also been featured a few times by National Geographic.

See also

- List of Islands in Palau
- List of Presidents of Palau

References

[1] CIA Factbook (https://www.cia.gov/library/publications/the-world-factbook/geos/ps.html)

[2] "Palau" (https://www.cia.gov/library/publications/the-world-factbook/geos/ps.html). *CIA World Factbook*. CIA. . Retrieved 2009-08-09.

[3] http://hdr.undp.org/en/media/HDR_2011_ES_Table1.pdf

[4] Nelson's World Gazetteer and Geographical Dictionary, ed. J. Gunn, Thomas Nelson and Sons Ltd, London, 1935

[5] *Compact of Free Association: Palau's Use of and Accountability for U.S. Assistance and Prospects for Economic Self-Sufficiency* (http://www.gao.gov/products/GAO-08-732). United States Government Accountability Office. June 10, 2008. p. 7.

[6] Compact of Free Association Between the Government of the United States of America and the government of Palau (http://palau.usembassy.gov/rop_cofa.pdf), Preamble

[7] Compact of Free Association Between the Government of the United States of America and the government of Palau (http://palau.usembassy.gov/rop_cofa.pdf), Table of Contents

[8] USDOI Office of Insular Affairs (http://www.doi.gov/oia/Firstpginfo/oia_responsibilities.html)

[9] "Pacific Nations Extend Fishing Ban" (http://pidp.eastwestcenter.org/pireport/2010/October/10-06-01.htm). *Radio Australia* (East West Center). 2010-10-05. . Retrieved 2010-10-06.

[10] "The Constitution of the Republic of Palau" (http://www.paclii.org/pw/legis/consol_act/cotrop359/). The Government of Palau. 2 April 1979. . Retrieved 1 November 2009.

[11] "Issues Associated. With Palau's Transition to Self-Government" (http://archive.gao.gov/d26t7/139356.pdf). Government Accountability Office. July 1989. . Retrieved 1 November 2009.

[12] Lyons, Richard D. (6 November 1994). "Work Ended, Trusteeship Council Resists U.N. Ax for Now" (http://www.nytimes.com/1994/11/06/world/work-ended-trusteeship-council-resists-un-ax-for-now.html?scp=2&sq=palau&st=nyt). New York Times. . Retrieved 1 November 2009.

[13] *Trusteeship Mission reports on Palau voting. (plebiscite on the Compact of Free Association with the United States).*. **27**. UN Chronicle. June 1990.

[14] Matthew Lee, Devlin Barrett (2009-06-09). "US eyes Pacific to resettle Uighur detainees" (http://www.webcitation.org/query?url=http://wtop.com/?nid=116&sid=1692515&date=2009-06-11). WTOP. Archived from the original (http://wtop.com/?nid=116&sid=1692515) on 2009-06-11. . Retrieved 2009-06-11.

[15] Ray Lilley (2009-06-10). "Island nation of Palau to take Gitmo's Uighurs" (http://www.webcitation.org/query?url=http://www.thestar.com/news/world/article/648438&date=2009-06-11). Toronto Star. Archived from the original (http://www.thestar.com/news/world/article/648438) on 2009-06-11. . Retrieved 2009-06-11.

[16] "Pacific state Palau to take Uighur detainees" (http://www.webcitation.org/query?url=http://www.ctv.ca/servlet/ArticleNews/story/CTVNews/20090610/Palau_Uighur_090610/20090610?hub=World&date=2009-06-11). CTV News. 2009-06-10. Archived from the original (http://www.ctv.ca/servlet/ArticleNews/story/CTVNews/20090610/Palau_Uighur_090610/20090610?hub=World) on 2009-06-11. . Retrieved 2009-06-11.

[17] "Palau to take 17 Uygur Guantanamo inmates" (http://www.webcitation.org/query?url=http://www.chinadaily.com.cn/world/2009-06/11/content_8272318.htm&date=2009-06-11). China Daily. 2009-06-11. Archived from the original (http://www.chinadaily.com.cn/world/2009-06/11/content_8272318.htm) on 2009-06-11. . Retrieved 2009-06-11.

[18] Mike Levine (2009-06-09). "Palau Agrees to Take Uighur Gitmo Detainees" (http://www.webcitation.org/query?url=http://www.foxnews.com/politics/2009/06/09/report-obama-administration-talks-palau-resettle-uighur-detainees/&date=2009-06-11). Fox News. Archived from the original (http://www.foxnews.com/politics/2009/06/09/report-obama-administration-talks-palau-resettle-uighur-detainees/) on 2009-06-11. . Retrieved 2009-06-11.

[19] "Palau to take Uighur detainees from Guantanamo Bay" (http://www.webcitation.org/query?url=http://news.yahoo.com/s/ap/20090610/ap_on_re_au_an/as_us_guantanamo_palau&date=2009-06-11). Yahoo News. 2009-06-10. Archived from the original (http://news.yahoo.com/s/ap/20090610/ap_on_re_au_an/as_us_guantanamo_palau) on 2009-06-11. . Retrieved 2009-06-11.

[20] Stuart Beck (2009-06-19). "The Myth of the $12 Million Uighur" (http://www.webcitation.org/query?url=http://www.nytimes.com/2009/06/20/opinion/20beck.html?em&date=2009-11-01). New York Times. Archived from the original (http://www.nytimes.com/2009/06/20/opinion/20beck.html?em) on 2009-11-01. .

[21] Kirit Radia (2009-06-10). "US and Palau wrangling over Gitmo transfer details, including $$$" (http://www.webcitation.org/query?url=http://blogs.abcnews.com/politicalpunch/2009/06/us-and-palau-wrangling-over-gitmo-transfer-details-including.html&date=2009-11-01). ABC News. Archived from the original (http://blogs.abcnews.com/politicalpunch/2009/06/us-and-palau-wrangling-over-gitmo-transfer-details-including.html) on 2009-11-01. .

[22] William Cleary (2009-06-24). "Will taking detainees pay off for Palau?" (http://www.guampdn.com/article/20090624/OPINION02/906240322/1014/OPINION/Will+taking+detainees+pay+off+for+Palau). Pacific Daily News. . Retrieved 2009-06-26.

[23] Kevin Kerrigan (2009-06-23). "Uyghures Don't Show Much Interest In Palau" (http://www.pacificnewscenter.com/index.php?option=com_content&view=article&id=8802:uyghures-dont-show-much-interest-in-palau&catid=34:guam&Itemid=141). Pacific News Center. . Retrieved 2009-06-28.

[24] "Ministry of State: Press Release" (http://www.webcitation.org/query?url=http://www.pacificnewscenter.com/images/pdfs/palauwegers.pdf&date=2009-11-01). Government of Palau. 2009-06-23. Archived from the original (http://www.pacificnewscenter.com/images/pdfs/palauwegers.pdf) on 2009-11-01. .

[25] "Palau Government still not sure if Uighurs are coming" (http://www.webcitation.org/query?url=http://www.rnzi.com/pages/news.php?op=read&id=47454&date=2009-11-01). Radio New Zealand International. 2009-06-30. Archived from the original (http://www.rnzi.com/pages/news.php?op=read&id=47454) on 2009-11-01. .

[26] Anne Barrowclough (2009-09-10). "Palau move imminent for Guantanamo Uighurs" (http://www.webcitation.org/query?url=http://www.timesonline.co.uk/tol/news/world/us_and_americas/article6828668.ece&date=2009-09-19). London: The Times. Archived from the original (http://www.timesonline.co.uk/tol/news/world/us_and_americas/article6828668.ece) on 2009-09-19. .

[27] "Six Guantanamo Uighurs arrive in Palau: US" (http://www.webcitation.org/query?url=http://www.google.com/hostednews/afp/article/ALeqM5j256OdOnArVNzoV0oVLD3PRaY-bQ&date=2009-10-31). Agence France Presse. 2009-10-31. Archived from the original (http://www.google.com/hostednews/afp/article/ALeqM5j256OdOnArVNzoV0oVLD3PRaY-bQ) on 2009-10-31. .

[28] "6 Muslim Uighur Detainees From Guantanamo Arrive In Palau" (http://www.webcitation.org/query?url=http://www.pacificnewscenter.com/index.php?option=com_content&view=article&id=1343:6-muslim-uighurs-arrive-in-palau-from-guantanamo&catid=45:guam-news&Itemid=156&date=2009-10-31). Pacific News Center. 2009-11-01. Archived from the original (http://www.pacificnewscenter.com/index.php?option=com_content&view=article&id=1343:6-muslim-uighurs-arrive-in-palau-from-guantanamo&catid=45:guam-news&Itemid=156) on 2009-10-31. . Retrieved 2009-09-26.

[29] "United States Transfers Six Uighur Detainees from Guantanamo Bay to Palau" (http://www.webcitation.org/query?url=http://www.justice.gov/opa/pr/2009/October/09-ag-1179.html&date=2009-10-31). United States Department of Justice. 2009-10-31. Archived from the original (http://www.justice.gov/opa/pr/2009/October/09-ag-1179.html) on 2009-10-31. .

[30] David Johnston (2009-10-31). "Uighurs Leave Guantánamo for Palau" (http://www.webcitation.org/query?url=http://www.nytimes.com/2009/11/01/world/asia/01uighurs.html&date=2009-10-31). New York Times. Archived from the original (http://www.nytimes.com/2009/11/01/world/asia/01uighurs.html) on 2009-10-31. .

[31] "Guantanamo Uighurs sent to Palau" (http://www.webcitation.org/query?url=http://news.bbc.co.uk/1/hi/world/asia-pacific/8336343.stm&date=2009-10-31). BBC News. 2009-10-31. Archived from the original (http://news.bbc.co.uk/1/hi/world/asia-pacific/8336343.stm) on 2009-10-31. .

[32] "Palau receives aid boost from US" (http://www.webcitation.org/query?url=http://australianetworknews.com/stories/201001/2804280.htm?desktop&date=2010-02-04). Australia Network News. 2010-01-30. Archived from the original (http://australianetworknews.com/stories/201001/2804280.htm?desktop) on 2010-02-04. . "The president insisted there was no link to the island's hosting of six inmates from Guantanamo Bay. Palau had earlier rejected a 156 million dollar offer and the settlement came after the island nation agreed to resettle six Muslim Uighurs who had been held for more than seven years at the US naval base at Guantanamo Bay. The six arrived in Palau in November. But Johnson said the two issues were not related."

[33] "Palau creates world's first shark haven" (http://www.philstar.com/Article.aspx?articleId=508795&publicationSubCategoryId=200). *The Philippine Star*. 2009-09-26. . Retrieved 2009-09-28.

[34] Black, Richard. "Palau pioneers 'shark sanctuary'. BBC News website. 25 September 2009. http://usproxy.bbc.com/2/hi/science/nature/8272508.stm.Retrieved 25 September 2009.

[35] "Palau's EEZ becomes shark sanctuary" (http://news.xinhuanet.com/english/2009-09/27/content_12117618.htm). Xinhua News Agency. 2009-09-27. . Retrieved 2009-09-28.

[36] Sophie Tedmanson (2009-09-26). "World's first shark sanctuary created by Pacific island of Palau" (http://www.timesonline.co.uk/tol/news/environment/article6848627.ece). London: *The Times*. . Retrieved 2009-09-28.

[37] Ker Than (2009-09-25). "France-Size Shark Sanctuary Created -- A First" (http://news.nationalgeographic.com/news/2009/09/090925-sharks-sanctuary-palau.html). *National Geographic*. . Retrieved 2009-09-28.

[38] "Palau creates shark sanctuary to protect tourism and prevent overfishing" (http://www.rnzi.com/pages/news.php?op=read&id=49344). Radio New Zealand. 2009-09-27. . Retrieved 2009-09-28.

[39] Cornelia Dean (2009-09-24). "Palau to Ban Shark Fishing" (http://www.nytimes.com/2009/09/25/science/earth/25shark.html). *The New York Times*. . Retrieved 2009-09-28.

[40] "Palau's PacificFlier relooks business plan after suspension" (http://www.flightglobal.com/articles/2010/08/13/346121/palaus-pacificflier-relooks-business-plan-after-suspension.html). . Retrieved September 13, 2011.

[41] "Taiwan pay for the bill of compensation for PSB bankrupcy!" (http://tw.news.yahoo.com/article/url/d/a/101220/2/2jatl.html). udn.com Center. . Retrieved 2010-12-20.

[42] "CIA - The World Factbook -- Field Listing :: Languages" (http://www.webcitation.org/5ncBXKnx7). Central Intelligence Agency. Archived from the original (https://www.cia.gov/library/publications/the-world-factbook/fields/2098.html) on February 17, 2010. . Retrieved February 17, 2010.

[43] Lewis, Paul M. (ed) (2009). "Languages of Palau" (http://www.webcitation.org/5ncBmM3tC). SIL International. Archived from the original (http://www.ethnologue.com/show_country.asp?name=Palau) on February 17, 2010. . Retrieved February 17, 2010.

[44] Office of Planning and Statistics, Republic of Palau: 2005 Census of Population and Housing (http://www.spc.int/prism/country/pw/stats/PalauStats/Publication/publications.htm), Table 59, retrieved on March 1, 2007

[45] Lawyers: Uighurs agree to go to Palau (http://www.google.com/hostednews/ap/article/ALeqM5h0PvMT3xGXLwI-2KwkOThMJToprwD9AK6D1O0)

[46] Lawyers: 3 Uighurs agree to go to Palau (http://www.google.com/hostednews/ap/article/ALeqM5h0PvMT3xGXLwI-2KwkOThMJToprwD9AK5OB81)

External links

- Palau (https://www.cia.gov/library/publications/the-world-factbook/geos/ps.html) entry at *The World Factbook*
- Palau (http://ucblibraries.colorado.edu/govpubs/for/Palau.htm) from the *University of Colorado at Boulder Libraries (USA) - Government Publications.*
- Australian library online (http://librariesaustralia.nla.gov.au/apps/kss), subscription, or membership via institutional access, is required.
- Palau (http://www.dmoz.org/Regional/Oceania/Palau/) at the Open Directory Project
- Wikimedia Atlas of Palau
- Palau travel guide from Wikitravel

Government

- Official Site of the Republic of Palau (http://www.palaugov.net)
- Palau National Congress - Olbiil Era Kelulau, Senate (http://senate.palauoek.net)
- Chief of State and Cabinet Members (https://www.cia.gov/library/publications/world-leaders-1/world-leaders-p/palau.html)

bjn:Palau mrj:Палау

Geography_of_Palau

The Republic of Palau consists of eight principal islands and more than 250 smaller ones lying roughly 500 miles southeast of the Philippines. The islands of Palau constitute part of the Caroline Islands chain.

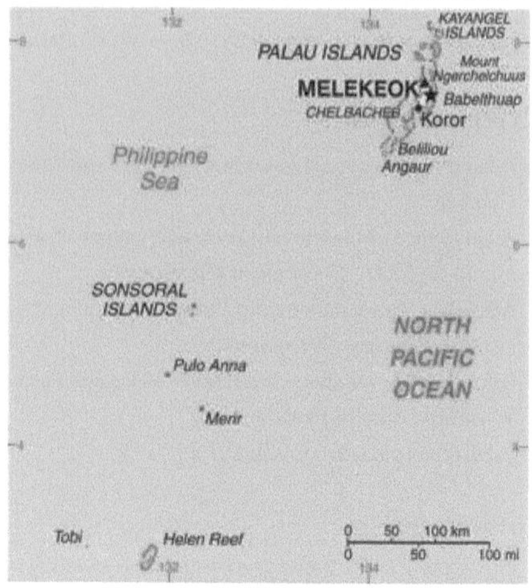

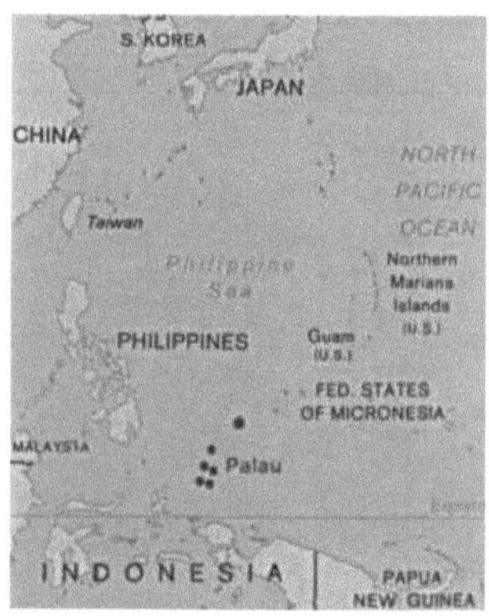

Map of Palau and its location in Oceania

Location: Oceania, group of islands in the North Pacific Ocean, southeast of the Philippines

Geographic coordinates: 7°30′N 134°30′E

Map references: Oceania

Area:
total: 458 km²
land: 458 km²
water: 0 km²

Area - comparative: slightly more than 2.5 times the size of Washington, DC

Land boundaries: 0 km

Coastline: 1,519 km

Maritime claims:
continental shelf: 200-m depth or to the depth of exploitation
exclusive fishing zone: 12 nm
extended fishing zone: 200 nm
territorial sea: 3 nm

Climate: wet season May to November; hot and humid

Terrain: varying geologically from the high, mountainous main island of Babeldaob to low, coral islands usually fringed by large barrier reefs

Elevation extremes:
lowest point: Pacific Ocean 0 m
highest point: Mount Ngerchelchauus 242 m (on Babeldaob)

Natural resources: forests, minerals (especially gold), marine products, deep-seabed minerals

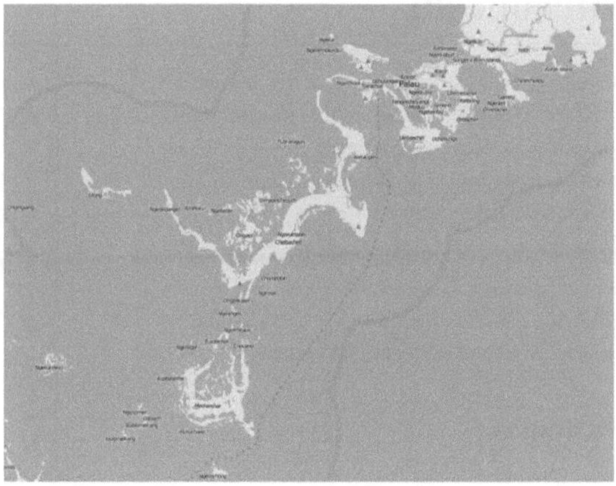

Land use:
arable land: NA%
permanent crops: NA%
permanent pastures: NA%
forests and woodland: NA%
other: NA%

Irrigated land: NA km²

Natural hazards: typhoons (June to December)

Environment - current issues: inadequate facilities for disposal of solid waste; threats to the marine ecosystem from sand and coral dredging, illegal fishing practices, and overfishing

Environment - international agreements:
party to: Biodiversity, Climate Change, Climate Change-Kyoto Protocol, Desertification, Law of the Sea
signed, but not ratified: none of the selected agreements

Geography - note: includes World War II battleground of Peleliu and world-famous rock islands; archipelago of six island groups totaling over 200 islands in the Caroline chain

Extreme points

This is a list of the extreme points of Palau, the points that are farther north, south, east or west than any other location.

- Northern-most point – Ngaruangel Reef, Kayangel state*
- Eastern-most point – Kayangel islet, Kayangel state
- Southern-most point – Round Rock, Helen Reef, Hatohobei state
- Western-most point - Tobi Island, Hatohobei state
- **Note: if submerged reefs are included then Velasco reef is Palau's northern-most point*

See also : Palau

External links

- Map of main island chain of Palau [1]

Micronesia

Micronesia is a subregion of Oceania, comprising thousands of small islands in the western Pacific Ocean. It is distinct from Melanesia to the south, and Polynesia to the east. The Philippines lie to the west, and Indonesia to the southwest.

The name Micronesia derives from the Greek *mikros* (μικρός), meaning *small*, and *nesos* (νῆσος), meaning *island*. The term was first proposed to distinguish the region in 1831 by Jules Dumont d'Urville.

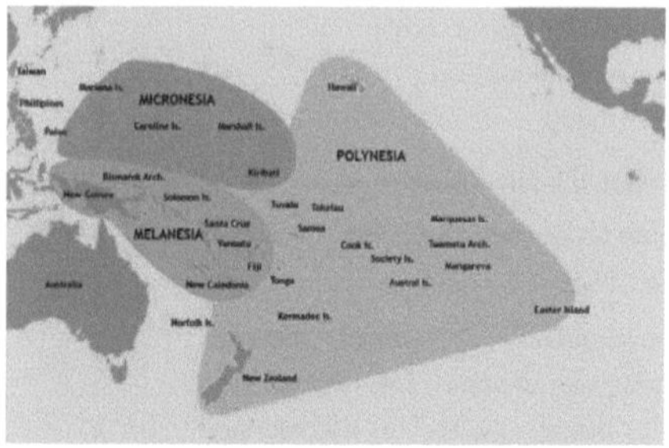

Micronesia is one of three major cultural areas in the Pacific Ocean.

Geography

Associated islands

The following islands and groups of islands are considered part of Micronesia:

- Banaba, an outlier of Kiribati
- Gilbert Islands, which forms part of Kiribati
- Mariana Islands, politically divided between Guam and the Northern Mariana Islands
- Marshall Islands
- Caroline Islands, politically divided between Palau and the Federated States of Micronesia
- Nauru
- Wake Island, a United States Minor Outlying Island

Total Land Area: 1229.95 square miles (3185.6 km^2)

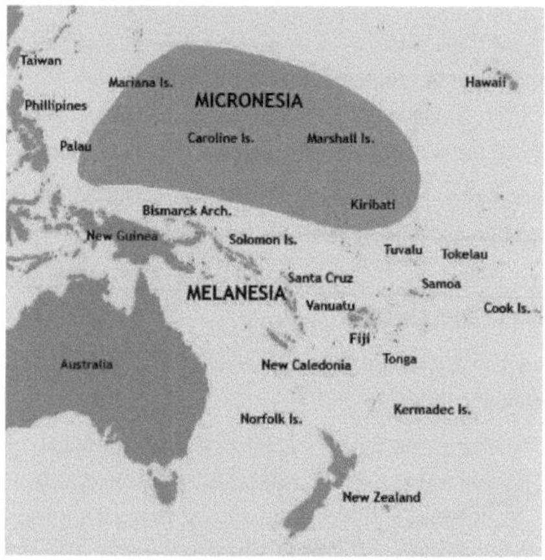

Map of Micronesia

History

The only empire known to have originated in Micronesia was based in Yap.

Much of the area came under European domination quite early. In the early 17th Century Spain colonized Guam, the Northern Marianas, and the Caroline Islands (what would later become the Federated States of Micronesia and Palau), creating the Spanish East Indies, which was governed from the Spanish Philippines until the Spanish-American War in 1898.

Full European colonization did not come, however, until the early 20th century, when the area would be divided between:

- the United States, which took control of Guam following the Spanish-American War of 1898, and colonized Wake Island;
- Germany, which took Nauru and bought the Marshall, Caroline, and Northern Mariana Islands from Spain; and
- the British Empire, which took the Gilbert Islands (Kiribati).

During the World War I, Germany's Pacific island territories were seized and they became League of Nations Mandates in 1923. Nauru became an Australian mandate, while Germany's other territories in Micronesia were given as a mandate to Japan and were named the South Pacific Mandate. Following Japan's defeat in the Second World War, its mandate became a United Nations Trusteeship, the Trust Territory of the Pacific Islands, ruled by the United States.

Today, most of Micronesia are independent states, except for Guam and Wake Island, which are U.S. territories, and for the Northern Mariana Islands, which are a U.S. commonwealth.

People

The people today form many ethnicities, but are all descended from and belong to the Micronesian culture. The Micronesian culture was one of the last native cultures of the region to develop. It developed from a mixture of Melanesians, Polynesians, and Filipinos. Because of this mixture of descent, many of the ethnicities of Micronesia feel closer to some groups in Melanesia, Polynesia or the Philippines. A good example of this are the Yapese who are related to Austronesian tribes in the Northern Philippines.[1]

Languages

The native languages of the various Micronesian indigenous peoples are classified under the Austronesian language family. Almost all of these languages belong to the Oceanic branch of this family, and mostly to the Micronesian languages division within that branch; however, two exceptions are noted in western Micronesia, which are Western Malayo-Polynesian languages:

- Chamorro in the Mariana Islands
- Palauan in Palau.

These are apparently no more closely related to Micronesian languages than they are to languages of the Philippines, Malaysia, and Indonesia (Kirch, 2000: pp. 166–167).

On the eastern edge of the Federated States of Micronesia, the languages Nukuoro and Kapingamarangi represent an extreme westward extension of Polynesian.

Regional organizations

The region is home to the Micronesian Games, a quadrennial international multi-sport event involving all Micronesia's countries and territories except Wake Island.

In September 2007, journalists in the region founded the Micronesian Media Association.[2]

See also

- List of mammals of Micronesia
- Melanesia
- Polynesia

References

[1] http://www.everyculture.com/wc/Mauritania-to-Nigeria/Micronesians.html
[2] Regional journalists form Micronesian media group (http://www.saipantribune.com/newsstory.aspx?newsID=72699&cat=1), *Saipan Tribune*, 26 September 2007

Further reading

- Kirch, Patrick Vinton (2000). *On the Road of the Winds. An Archaeological History of the Pacific Islands before European Contact.* University of California Press. pp. 166–167. ISBN 0-520-22347-0.
- Goetzfridt, Nicholas J. and Karen M. Peacock. (2002) Micronesian Histories: An Analytical Bibliography and Guide to Interpretations. Westport. CT: Greenwood Press.

External links

- History of Micronesia (http://www.visit-fsm.org/visitors/history.html)
- Micronesian Games (http://www.gbrathletics.com/ic/micg.htm)

Island_country

An **island country** is a state whose primary territory consists of one or more islands or parts of islands. As of 2011, 47 (approximately 25%[1]) of the 193 UN member states are island countries.

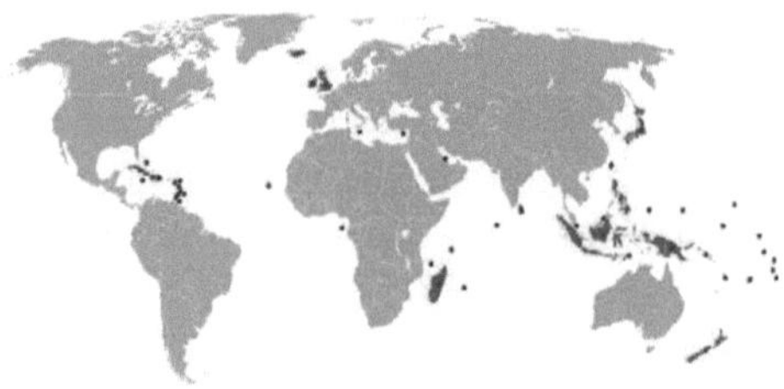

Island countries in the world

Politics

The percentage of island countries that are democratic is higher than that of continental countries, although historically they have been more prone to political instability.[1]

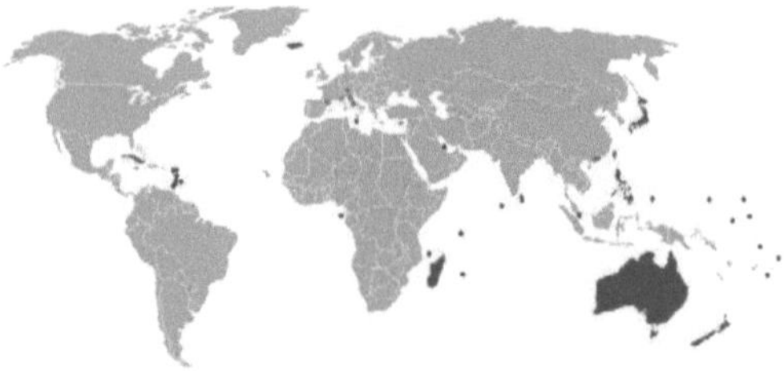

Countries without land borders

War

Island countries have often been the basis of maritime conquest and historical rivalry between other countries.[2] Island countries are more susceptible to attack by large, continental countries due to their size and dependence on sea lines of communication and air lines of communication.[3] Many island countries are also vulnerable to predation by mercenaries and other foreign invaders.[4] However, their isolation also makes them a difficult target.

Natural resources

Many island countries rely heavily on fish for their main supply of food.[5] Some are turning to renewable energy—such as wind power, hydropower, geothermal power and biodiesel from copra oil—to combat the rise in oil prices.[6]

Geography

Some island countries are more affected by climate change than other countries. Some low-lying island countries are slowly being submerged by the rising water levels of the Pacific Ocean.[7] Climate change also impacts island countries by causing natural disasters such as tropical cyclones, hurricanes, flash floods and drought.[8]

Economics

Many island countries rely heavily on imports and are greatly affected by changes in the global economy.[9] The economies of island countries are usually smaller and more vulnerable to shipping costs, environmental damage to infrastructure, and isolation from other economies than those of larger, continental countries (excluding Japan, Australia and the U.K.) [10] [11] [12] The dominant industry for many island countries is tourism.[13]

Composition

Island countries are typically small with relatively low populations.[14]

Some island countries are centred on one or two major islands, such as the United Kingdom, Trinidad and Tobago, New Zealand and Singapore. Others are spread out over hundreds or thousands of smaller islands, such as the Philippines, Indonesia, and the Maldives. Some island countries share their islands with other countries; these include the United Kingdom and the Republic of Ireland, Haiti and the Dominican Republic, and Indonesia, which shares islands with Papua New Guinea, Brunei, East Timor and Malaysia.

Geographically, the country of Australia is considered a continental landmass rather than an island, covering the largest landmass of the Australian continent. In the past, however, it was considered an island country for tourism purposes[15] and is still often referred to as such.[16]

Climate change

Climate change poses a threat to island nations. Rising sea levels provide obstacles including reduced land use, water scarcity and sometimes even resettlement issues for island nations. In 2011, the Center for Climate Change Law (CCCL) held a conference attended by 272 registrants from 39 island nations titled Legal Issues for Threatened Island Nations.[17]

See also

- Archipelago
- List of Caribbean island countries by population
- List of countries
- List of island countries
- List of islands by area
- List of islands (by country)
- Micronation
- Pacific Islands Forum
- Small Island Developing States

References

[1] Ott, Dana (2000). *Small is Democratic* (http://books.google.com/books?id=JOqZ9eOeRXoC&pg=PA128&dq="island+country"&lr=#PPA128,M1). Routledge. pp. 128. ISBN 0815339100. .

[2] Chasle, Raymond (01-OCT-1986). "The quest for identity. (island countries)" (http://www.accessmylibrary.com/coms2/summary_0286-9197350_ITM). UNESCO Courier. . Retrieved 2009-02-01.

[3] Perry, Walt L.; Robert W. Button, Jerome Bracken, Thomas Sullivan, Rand Corporation, United States Navy, Jonathan Mitchell (2002). *Measures of Effectiveness for the Information-age Navy* (http://books.google.com/books?id=VwI0WLby3WoC&pg=PA7&dq="island+country"&lr=#PPA7,M1). Rand Corporation. pp. 7. ISBN 0833031392. .

[4] WREN, CHRISTOPHER S. (December 9, 1989). "Mercenary Holding Island Nation Seeks Deal" (http://query.nytimes.com/gst/fullpage.html?res=950DE3DB173CF93AA35751C1A96F948260). New York Times. . Retrieved 2009-02-01.

[5] "OTTAWA: REPORT ISSUES WARNING ABOUT FUTURE OF CORAL FISHERIES" (http://www.rcinet.ca/rci/en/news/2007/04/20070421.shtml). Radio Canada International. 2007-04-21. . Retrieved 2009-02-01.

[6] Xingwei, Huang (2008-10-17). "Pacific Islands countries switch to renewable energy source due to increasing fuel prices" (http://news.xinhuanet.com/english/2008-10/17/content_10210370.htm). . Retrieved 2009-02-01.

[7] "Leader of disappearing island nation says climate change an issue of survival, not economics" (http://www.iht.com/articles/ap/2008/06/05/asia/AS-GEN-New-Zealand-World-Environment-Day.php). June 5, 2008. . Retrieved 2009-02-01.

[8] "FAO: Climate change threatens food security of Pacific island countries" (http://english.people.com.cn/90001/90777/90856/6544815.html). December 02, 2008. . Retrieved 2009-02-01.

[9] "Global economic crisis a concern for Pacific island countries" (http://www.cctv.com/program/bizchina/20081028/106856.shtml). 2008-10-28. . Retrieved 2009-02-01.

[10] "WTO/FORSEC Trade Policy Course for Pacific island countries" (http://www.wto.org/english/news_e/news01_e/fiji_e.htm). 9 March 2001. . Retrieved 2009-02-01.

[11] "NZ calls for global solutions to problems faced by small island nations" (http://news.xinhuanet.com/english/2005-01/18/content_2475179.htm). 2005-01-18. . Retrieved 2009-02-01.

[12] Garg, Sarika. "U.N. ambassador gives keynote" (http://www.dailypennsylvanian.com/home/index.cfm?event=displayArticlePrinterFriendly&uStory_id=ce109893-987e-42f8-816f-019cbd58eefa). . Retrieved 2009-02-01.

[13] "China enlists Pacific island countries as tourist destinations, XINHUA." (http://www.accessmylibrary.com/coms2/summary_0286-12915601_ITM). *The America's Intelligence Wire*. 10-AUG-04. . Retrieved 2009-02-01.

[14] "Wen pledges new aid to Pacific countries" (http://www.iht.com/articles/2006/04/05/news/pacific.php). International Herald Tribute. April 5, 2006. . Retrieved 2009-02-01.

[15] "Australian Naval Defence" (http://ndpbeta.nla.gov.au/ndp/del/article/3655029). The Brisbane Courier. 24 July 1897. . Retrieved 2009-01-31.

[16] "China, Pacific island countries discuss cooperation at forum meeting" (http://english.people.com.cn/200604/05/eng20060405_256086.html). . Retrieved 2009-02-01.

[17] http://www.earth.columbia.edu/articles/view/2826

Mount_Ngerchelchuus

Mount Ngerchelchuus	
 Mount Ngerchelchuus	
Elevation	242 m (794 ft) [1]
Prominence	242 m (794 ft) [1]
Listing	Country high point
Location	
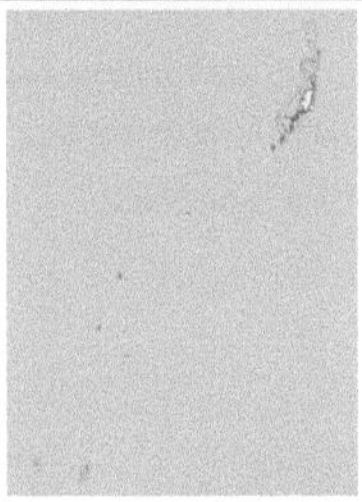 Mount Ngerchelchuus	
Coordinates	7°33′55″N 134°34′10″E

Mount Ngerchelchuus is the Republic of Palau's highest point, located at the border of the states of Ngardmau and Ngaremlengui, on the island of Babeldaob.

References

[1] "Mount Ngerchelchauus" on Peakbagger (http://www.peakbagger.com/peak.aspx?pid=11791) Retrieved 24 September 2011

Babeldaob

Babeldaob (also **Babelthuap**) is the largest island in the island nation of Palau. Its area, 331 km² (128 mi²), makes up over 70% of the area of the entire country, and about 30% of the population, with about 6,000 people living there. Babeldoab is one of the most undeveloped populated islands in the Pacific.

Beach near Melekeok

Geography

Babeldaob is located northeast of Koror and contains the site of the new national capital, Ngerulmud. Unlike most of the islands of Palau, Babeldaob is mountainous. It contains Palau's highest point, the 242 meter tall Mount Ngerchelchuus. Babeldaob's eastern coast has many sandy beaches, in particular north from Ngiwal to Ngaraard, and the island's western coast has a shoreline with many mangrove shrubs and trees.[1]

Palau's second-most populous state, Airai, makes up the southern end of the island. The Airai airport is the nation's principal airport. The Koror-Babeldaob Bridge links Airai to Koror.

Babeldaob encompasses 10 of the 16 states of Palau, which are:

- Aimeliik
- Airai
- Melekeok
- Ngaraard
- Ngarchelong
- Ngardmau
- Ngatpang
- Ngchesar
- Ngeremlengui
- Ngiwal

History

Babeldaob was the location to which 426 members of the Sokehs tribe were banished by the German Empire following the Sokehs Rebellion on Sokehs Island and Pohnpei. Japan as mandatory power after World War I administered the return of the Sokehs in stages between 1917 and 1927.[2]

During World War II the Japanese garrison was composed of 21,449 IJA men under the command of Lieutenant-General Sadae Inoue and 8,286 IJN men under the command of Vice-Admiral Kenmi Itoh.[3] Babeldaob was bypassed by the allied forces during the war.[3]

Notes

[1] Babeldaob information (http://www.lonelyplanet.com/palau/babeldaob) from Lonely Planet (http://www.lonelyplanet.com/). Retrieved June 2009.

[2] Micronesian Seminar: The Sokehs Rebellion (http://www.micsem.org/photos/sokehs/intro.htm)

[3] Takizawa, Akira; Alsleben, Allan (1999-2000). "Japanese garrisons on the by-passed Pacific Islands 1944-1945" (http://www.dutcheastindies.webs.com/japan_garrison.html). *Forgotten Campaign: The Dutch East Indies Campaign 1941-1942.* .

References

- L, Klemen (1999-2000). "Forgotten Campaign: The Dutch East Indies Campaign 1941-1942" (http://www.dutcheastindies.webs.com/index.html).

Environment_of_Palau

The **environment of Palau** consists a number of islands in the western Pacific Ocean.

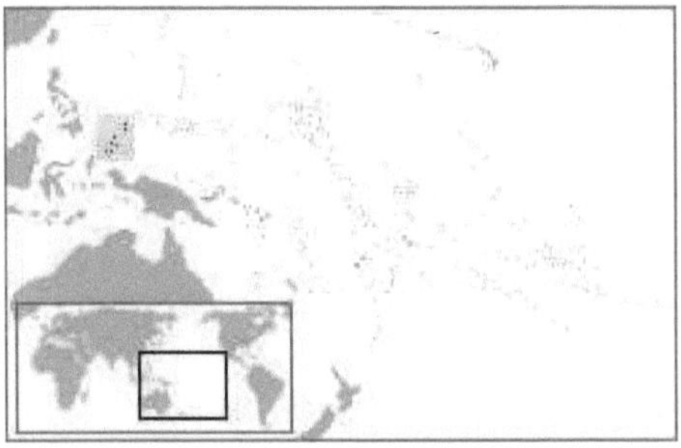

Palau is to the north of Australia and to the southeast of the Philippines.

Biota

The avifauna of Palau includes a total of 149 species, of which 10 are endemic, three have been introduced by humans, and 17 are rare or accidental. Three species are globally threatened.

Climate

The Palau government are concerned about the effects of climate change on the island nation. In 2008 Palau requested that the UN Security Council consider protection against rising sea levels due to climate change.[1]

Tommy Remengesau, the former president of Palau and now Senator, has said:

> Palau has lost at least one third of its coral reefs due to climate change related weather patterns. We also lost most of our agricultural production due to drought and extreme high tides. These are not theoretical, scientific losses -- they are the losses of our resources and our livelihoods.... For island states, time is not running out. It has run out. And our path may very well be the window to your own future and the future of our planet.

Geography

Palau consists of eight principal islands and more than 250 smaller ones lying roughly 500 miles southeast of the Philippines. The islands of Palau constitute part of the Caroline Islands chain.

See also

- List of environmental issues

References

[1] "Palau Calls for Security Council Action To Protect Island Nations From Sea-Level Rise" (http://www.palauun.org/news_archive.cfm?news_id=189). Permanent Representative of Palau to the United Nations. 2008-02-15. . Retrieved 2008-09-04.

External links

- Republic of Palau - National Government (http://www.palaugov.net/index.html)

Wildlife_of_Palau

For information on the **wildlife of Palau**, see:

- List of birds of Palau
- List of mammals of Palau

States_of_Palau

Palau is divided into sixteen administrative regions, called states (the still separately listed Rock Islands did originally not belong to any state, but now belong to the state of Koror).

State	Capital	Area (km^2)	Population (Census 2005)
North of Babeldaob			
Kayangel	Kayangel	3	188
Babeldaob			
Aimeliik	Mongami	52	270
Airai	Ngetkib	44	2 723
Melekeok	Melekeok	28	391
Ngaraard	Ulimang	36	581
Ngarchelong	Mengellang	10	488
Ngardmau	Urdmang	47	166
Ngaremlengui	Imeong	65	317
Ngatpang	Ngereklmadel	47	464
Ngchesar	Ngersuul	41	254
Ngiwal	Ngerkeai	26	223
Southwest of Babeldaob			

Angaur	Ngaramasch	8	320
Koror	Koror	18	12 676
Peleliu	Kloulklubed	13	702
Rock Islands, (also called *Chelbacheb*)	–	47	–
Southwest Islands			
Hatohobei	Hatohobei	3	44
Sonsorol	Dongosaru	3	100
Palau			
Palau	Ngerulmud[1]	488	19 907
[1] The national capital of Palau was moved from Koror to Ngerulmud on October 7, 2006.			

States, Municipalities, and Places

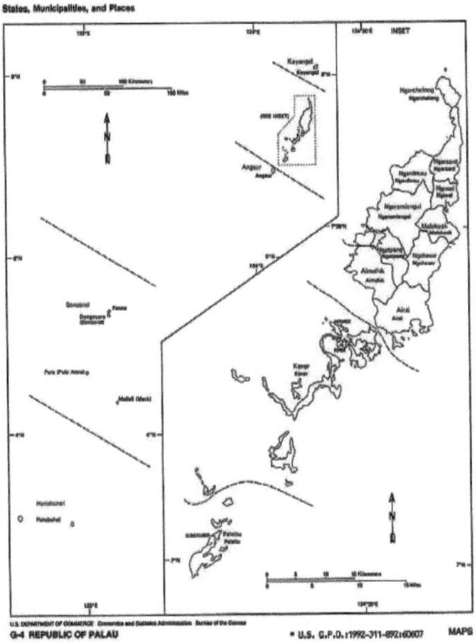

See also

- ISO 3166-2:PW

External links

- Republic of Palau – Palau's Sixteen States [1]
- Statoids [2]

mrj:Палау штатвлä

Demographics_of_Palau

This article is about the demographic features of the population of Palau, including population density, ethnicity, education level, health of the populace, economic status, religious affiliations and other aspects of the population.

About 70% of the Palauan population lives in the city of Koror on Koror Island. Koror is the former capital and the present capital is Melekeok on the bigger but less developed island of Babeldaob--the second-largest island in Micronesia after Guam.

CIA World Factbook demographic statistics

The following demographic statistics are from the CIA World Factbook, unless otherwise indicated.

Population

[1] 20,279

Age structure

0-14 years: 27% (male 2,605; female 2,458)
15-64 years: 68% (male 7,006; female 5,814)
65 years and over: 5% (male 416; female 467) (2000 est.)

Population growth rate

1.75% (2000 est.)

Birth rate

19.88 births/1,000 population (2000 est.)

Death rate

7.35 deaths/1,000 population (2000 est.)

Net migration rate

5.01 migrant(s)/1,000 population (2000 est.)

Sex ratio

at birth: 1.06 male(s)/female
under 15 years: 1.06 male(s)/female
15-64 years: 1.21 male(s)/female
65 years and over: 0.89 male(s)/female
total population: 1.15 male(s)/female (2000 est.)

Infant mortality rate

17.12 deaths/1,000 live births (2000 est.)

Life expectancy at birth

total population: 68.59 years
male: 65.47 years
female: 71.88 years (2000 est.)

Total fertility rate

2.47 children born/woman (2000 est.)

Nationality

noun: Palauan(s)
adjective: Palauan

Ethnic groups

Palauans are Micronesian with Melanesian and Malayan admixtures , Japanese , Filipino , Chinese , Micronesian(Chuukese,Phonpein,etc.)

Religions

[2]

(see Religion in Palau)

- Christian: 77.9%
  - Catholics 49.4%
  - Protestants 21.3%
  - Seventh-Day Adventists 5.3%
  - Jehovah's Witnesses 1.1%
  - Mormons 0.8%
- Modekngei religion (indigenous to Palau): 8.7%
- Other religion: 8.1%
- None or Refused: 1.1%

Languages

English and Palauan official in all states except Sonsoral (Sonsorolese and English are official), Hatohobei (Tobi and English are official), and Angaur (Japanese, and English are official)[3] [4]

Literacy

definition: age 15 and over can read and write
total population: 92%
male: 93%
female: 90% (1980 est.)

See also : Palau

References

[1] Office of Planning & Statistics: Census 2005 - Social statistics (http://www.spc.int/prism/country/pw/stats/PalauStats/Social/Census/Census.htm), retrieved on March 1, 2007

[2] Office of Planning and Statistics, Republic of Palau: 2005 Census of Population and Housing (http://www.spc.int/prism/country/pw/stats/PalauStats/Publication/publications.htm), Table 59, retrieved on March 1, 2007

[3] "CIA - The World Factbook -- Field Listing :: Languages" (http://www.webcitation.org/5ncBXKnx7). Central Intelligence Agency. Archived from the original (https://www.cia.gov/library/publications/the-world-factbook/fields/2098.html) on February 17, 2010. . Retrieved February 17, 2010.

[4] Lewis, Paul M. (ed) (2009). "Languages of Palau" (http://www.webcitation.org/5ncBmM3tC). SIL International. Archived from the original (http://www.ethnologue.com/show_country.asp?name=Palau) on February 17, 2010. . Retrieved February 17, 2010.

Melekeok

Melekeok is a state of the Republic of Palau located on the central east coast of *Babeldaob* Island. The new capital of the Republic of Palau, Ngerulmud, is also located in this state. The state consists of long beaches, hill, steep ridges, rivers and the largest and only natural freshwater lake in the Republic of Palau and Micronesia, Lake Ngardok.

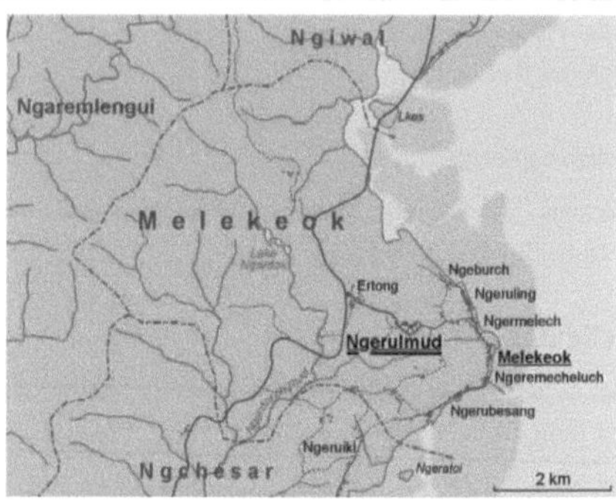

Map of State of Melekeok

Geography

Melekeok is located on the eastern coast of the largest island in the Palau Archipelago, *Babeldaob* or *Babelthuap* in some cases. Melekeok State extends from a lagoon, on the east coast, to the *Central Divide* or Rael Kedam to the west, and is approximately 11 square miles (28 km^2) in total area. The state is bordered by Ngiwal State to the north, Ngchesar State to the south, and Ngaremlengui State to the west. The state's coasts consist of long beaches and mangrove forests; the interior consists of swampy marshes, steep, rolling hills covered with sparse vegetation in the interior, and thick forests along the eastern flanks of the Rael Kedam Divide.

Melekeok's Lake Ngardok is the largest natural body of fresh water in all of Micronesia at 4.93 km^2; it has a small population of saltwater crocodiles *Crocodylus porosus*.

State Organizations

The State of Melekeok consists of eight inhabited villages: Ngeburch, Ngeruling, Ngermelech, Melekeok, Ngerubesang, Ngeremecheluch, Ertong, Ngerulmud. The state capitol is located in the settlement of Ngerulmud. The Republic of Palau's Capitol Complex is located on a hill, northwest of the village of Melekeok.

Demographics

According to the 2005 census, 391 people resided in the State of Melekeok. They are distributed among seven villages, including Melekeok, the capital:

1. Melekeok (271)
2. Ertong (20)
3. Ngeburch (34)
4. Ngeremecheluch (38)

5. Ngermelech (5)
6. Ngerubesang (12)
7. Ngeruling (21)

Registered Attractions

Odalmelech (Ngermelech Gods carved from Stone), Ngermelech Village; Meteu el Klechem, Melekeok Village; Did el Bad er a Beriber (Ancient Dock's stone bridge), Melekeok Village; Ngermecheluch Site, Ngerang; Bai el Melekeong/Euang el Chades(Melekeok's Bai and its platform of four), Melekeok Village; Lomech er Ngii el mora Ngeremecheluch (ancient stone path), Melekeok Village;

Flag of Melekeok

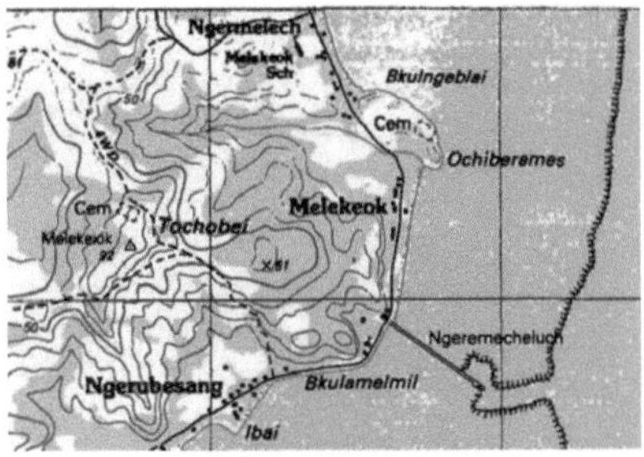

Melekeok area on topographic map

External links

- Seacology Melekeok State Project [1] Seacology
- Melekeok State [2]

Politics_of_Palau

Politics of Palau takes place in a framework of a presidential representative democratic republic, whereby the President of Palau is both head of state and head of government, and of a multi-party system. Executive power is exercised by the government. Legislative power is vested in both the government and the Palau National Congress. The Judiciary is independent of the executive and the legislature.

Executive branch

Presidential elections take place every four years, when the president and vice president run on separate tickets. The president, who is the chief of state and head of government, is currently Johnson Toribiong.

Palau Capitol Building

Legislative branch

The Palau National Congress (*Olbiil era Kelulau*) consists of two chambers—the Senate and the House of Delegates. The Senate has nine members elected nationwide. The House has 16 members, one from each of Palau's states. All legislators serve 4-year terms. Each state also elects its own governor and legislature.

The Council of Chiefs advises the president on traditional laws and customs. It is made up of the highest traditional chiefs from each state. Yutaka Gibbons (of Koror) is Chairman of the Council of Chiefs.

Political parties and elections

Summary of the 2 November 2004 Palau National Congress election results

Members	House of Delegates	Senate
Non-partisans	16	9
Total	16	9

Judicial branch

The judicial system consists of the Supreme Court, National Court, the Court of Common Pleas, and the Land Court. The Supreme Court has trial and appellate divisions and is presided over by the Chief Justice.

Political history

Palau adopted a constitution on January 1, 1981.

While calm in recent years, Palau witnessed several instances of political violence in the 1980s. The republic's first president, Haruo I. Remeliik, was assassinated in 1985; the Minister of State was found to be complicit in the crime. Palau's third president, Lazarus Salii, committed suicide in August 1988 amid bribery allegations. Salii's personal assistant had been imprisoned several months earlier after being convicted of firing shots into the home of the Speaker of the House of Delegates.

Palau gained independence from the United Nations trusteeship administered by the United States on 1 October 1994 and entered a Compact of Free Association with the United States.

The Senate passed legislation making Palau an "offshore" financial center in 1998. Opponents to the legislation voiced fears that the country would become a haven for money launderers and other sorts of criminal activity.

Oceania

WARNING: Article could not be rendered - ouputting plain text.
Potential causes of the problem are: (a) a bug in the pdf-writer software (b) problematic Mediawiki markup (c) table is too wide

An orthographic projection of the Pacific Ocean showing much of Oceania.Map of OceaniaOceania is a region centered on the islands of the tropical Pacific Ocean.For a history of the term, see Douglas & Ballard (2008) Foreign bodies: Oceania and the science of race 1750–1940 Conceptions of what constitutes Oceania range from the coral atolls and volcanic islands of the Pacific OceanSouth Pacific (Ethnologyethnologically divided into the subregions of Melanesia, Micronesia, and Polynesia "Oceania". 2005. The Columbia Encyclopedia, 6th ed. Columbia University Press.) to the entire insular region between Asia and the Americas, including Australasia and the Malay Archipelago." Oceania". Oxford English Dictionary. Oxford University Press. 2nd ed. 1989. The term is sometimes used more specifically to denote a continent comprising Australia (continent)Australia and proximate islands, Composition of macro geographical (continental) regions, geographical sub-regions, and selected economic and other groupings, United Nations Statistics Division. Revised August 28, 2007. Accessed on line October 11, 2007. The Atlas of Canada. Revised Date Modified: August 17, 2004. Accessed on line January 31, 2011. "Encarta Mexico "Oceanía"". Mx.encarta.msn.com. Archived from the original on 2009-11-01. . Retrieved 2009-04-17.Lewis, Martin W.; Kären E. Wigen (1997). The Myth of Continents: a Critique of Metageography. Berkeley: University of California Press. pp. 32. ISBN 0-520-20742-4, ISBN 0-520-20743-2. "Interestingly enough, the answer [from a scholar who sought to calculate the number of continents] conformed almost precisely to the conventional list: North America, South America, Europe, Asia, Oceania (Australia plus New Zealand), Africa, and Antarctica." or biogeographically as a synonym for either the Australasian ecozone (Wallacea and Australasia) or the Pacific ecozone (Melanesia, Polynesia, and Micronesia apart either from New ZealandUdvardy. 1975. A classification of the biogeographical provinces of the world or from mainland New GuineaSteadman. 2006. Extinction & biogeography of tropical Pacific birds). Etymology The term was coined as Océanie ca. 1812 by geographer Conrad Malte-Brun.

The word Océanie derives from the Greek languageGreek word ὠκεανός (ōkeanós), ocean.DefinitionsOceaniaAn orthographic projection of geopolitical Oceania.Geopolitical Oceania Demonym Oceanic; Oceanian Area 8536716 km2 (sq mi) Population 35,670,000 Countries Dependencies Languages Time Zones UTC+8 (Time in AustraliaAustralian Western Standard Time) to UTC-6 (Easter Island) (West to East) Largest Cities SydneyMelbourneBrisbanePerth, Western AustraliaPerthAucklandAdelaide, South AustraliaAdelaidePhysiography Oceania was originally conceived as the lands of the Pacific Ocean, stretching from the Straits of Malacca to the coast of the Americas. It comprised four regions: Polynesia, Micronesia, Malaysia (now called the Malay Archipelago), and Melanesia (now called Australasia).Jules-Sébastien-César Dumont D'UrvilleD'Urville, Jules-Sébastien-César Dumont; Isabel Ollivier, Antoine de Biran, and Geoffrey Clark. "On the Islands of the Great Ocean". The Journal of Pacific History (Taylor & Francis, Ltd.) 38 (2). . Included are parts of three geological continents, Eurasia, Australia (continent)Australia, and Zealandia, as well the non-continental volcanic islands of the Philippines, Wallacea, and the open Pacific. It extends to Sumatra in the west, the Bonin Islands in the northwest, the Hawaiian Islands in the northeast, Rapa Nui and Sala y Gómez Island in the east, and Macquarie Island in the south, but excludes Taiwan and the Ryukyu, Japanese, and Aleutian Islands of the margins of Asia.MacKay (1864, 1885) Elements of Modern Geography, p 283Douglas & Ballard (2008) Foreign bodies: Oceania and the science of race 1750–1940The states that occupy Oceania that are not included in geopolitical Oceania are Indonesia, Malaysia (through Malaysian Borneo), Brunei, the Philippines, and East Timor. The islands of the geographic extremes are politically integral parts of Japan (Bonin), the United States (Hawaii), and Chile (Easter Island). A smaller geographic definition also exists, which excludes the land on the Sunda Plate, but includes Indonesian New Guinea as part of the Australia (continent)Australian continent. Biogeography Biogeographically, Oceania is used as a synonym for either the Australasian ecozone (Wallacea and Australasia) or the Pacific ecozone (Melanesia, Polynesia, and Micronesia apart either from New Zealand or from mainland New Guinea). Ecogeography Oceania is one of eight terrestrial ecozones, which constitute the major ecological regions of the planet. The Oceania ecozone includes all of Micronesia, Fiji, and all of Polynesia except New Zealand. New Zealand, New Guinea, Melanesia apart from Fiji, and Australia constitute the separate Australasia ecozone. The Malay Archipelago is part of the Indomalaya ecozone. Related to these concepts are Near Oceania, that part of western Island Melanesia which has been inhabited for tens of millennia, and Remote Oceania, which is more recently settled.Ben Finney, The Other One-Third of the Globe, Journal of World History, Vol. 5, No. 2, Fall, 1994.GeopoliticsEconomic zones of the Pacific, outlining OceaniaIn the geopolitical conception used by the United Nations geoschemeUnited Nations, International Olympic Committee, and many atlases, Oceania includes Australia and the nations of the Pacific from Papua New Guinea east, but not the Malay Archipelago or Indonesian New Guinea. "United Nations Statistics Division - Countries of Oceania". Millenniumindicators.un.org. . Retrieved 2009-04-17.Atlas of Canada Web Master (2004-08-17). "The Atlas of Canada - The World - Continents". Atlas.nrcan.gc.ca. . Retrieved 2009-04-17. Current IOC members.Other definitions The term is sometimes used more specifically than in the geopolitical conception, to denote a continent comprising Australia (continent)Australia and proximate islands. New Zealand forms the south-western corner of the Polynesian Triangle. Its indigenous Māori peopleMāori constitute one of the major cultures of Polynesia. It is also, however, considered part of Australasia. The widest definition of Oceania includes the entire region between continental Asia and the Americas, thereby including islands in the Pacific Rim such as the Japanese Archipelago, Taiwan, and the Aleutian islands. Britannica Online EncyclopediaSatellite image of OceaniaEthno-cultural definition of OceaniaDemographicsOceaniaAn orthographic projection of geographic Oceania.Wider Geographic Oceania. Little of the South Pacific is apparent at this scale, though Hawaii is just visible near the eastern horizon. Area 10975600 km2 (sq mi) Population 378 million (2010) Time Zones UTC+7 (Time in IndonesiaWestern Indonesian Time) to UTC-6 (Easter Island) Largest Cities JakartaManilaSydneyBandungMelbourneSurabayaMedanNarrower Geographic Oceania.Narrower Geographic Oceania. Island Melanesia, Micronesia, and Polynesia apart from New Zealand) Area 183000 km2 (71000 sq mi) Population 5.2 million (2008) Time Zones UTC+9 (Palau) to UTC-6 (Easter Island) Largest Cities HonoluluNouméaSuvaPapeeteHoniaraGeographic map of islands of OceaniaThe demographic table below shows the subregions and countries of geopolitical Oceania. The countries and territories in this table are categorized according to the scheme for geographic subregions used by the United Nations. The information shown follows sources in cross-referenced articles; where sources differ, provisos have been clearly indicated. These territories and regions are subject to various additional categorisations, of course, depending on the source and purpose of each description. Name of region, followed by countriesand their flagsRegions and constituents as per UN categorisations/map except #endnote CCAUnotes 2-3, 6. Depending on definitions, various territories cited below (notes 3, 5-7, 9) may be in List of countries spanning more than one continentone or both of Oceania and Asia or North America.List of countries and outlying territories by total areaArea(km²) List of countries by populationPopulationList of countries and dependencies by population densityPopulation density(per km²) Capital (political)CapitalISO 3166-1 alpha-2ISO 3166-1AustralasiaThe use and scope of this term varies. The UN designation for this subregion is "Australia and New Zealand." AustraliaAustralia 7,686,850 22,028,000 2.7 Canberra AU New ZealandNew ZealandNew Zealand is often considered part of Polynesia rather than Australasia. 268,680 4,108,037 14.5 Wellington NZ External territories of Australia: Ashmore and Cartier Islands 199 Christmas IslandChristmas IslandChristmas Island and Cocos (Keeling) Islands are States and territories of AustraliaAustralian external territories in the Indian Ocean southwest of Indonesia. 135 1,493 3.5 Flying Fish Cove CX Cocos (Keeling) IslandsCocos (Keeling) Islands 14 628 45.1 West Island, Cocos (Keeling) IslandsWest Island CC Coral Sea Islands 10 4 Heard Island and McDonald Islands 372 Norfolk IslandNorfolk Island 35 2,114 53.3 Kingston, Norfolk IslandKingston NF MelanesiaExcludes parts of Indonesia, island territories in Southeast Asia (UN region) frequently reckoned in this region. FijiFiji 18,270 856,346 46.9 Suva FJ New CaledoniaNew Caledonia (France) 19,060 240,390 12.6 Nouméa NC Papua New GuineaPapua New GuineaPapua New Guinea is often considered part of Australasia and Melanesia. It is sometimes included in the Malay Archipelago of Southeast Asia. 462,840 5,172,033 11.2 Port Moresby PG Solomon IslandsSolomon Islands 28,450 494,786 17.4 Honiara SB VanuatuVanuatu 12,200 240,000 19.7 Port Vila VU Micronesia Federated States of MicronesiaFederated States of Micronesia 702 135,869 193.5 Palikir FM GuamGuam (United StatesUSA) 549 160,796 292.9 Hagåtña, GuamHagåtña GU KiribatiKiribati 811 96,335 118.8 South Tarawa KI Marshall IslandsMarshall Islands 181 73,630 406.8 Majuro MH NauruNauru 21 12,329 587.1 Yaren DistrictYaren (de facto) NR Northern Mariana IslandsNorthern Mariana Islands (USA) 477 77,311 162.1 Saipan MP PalauPalau 458 19,409 42.4 MelekeokOn 7 October 2006, government officials moved their offices in the former capital of Koror to Melekeok, located 20 km northeast of Koror on BabeldaobBabelthuap Island. PW Wake Island (USA) 2 12 Wake Island UM Polynesia American SamoaAmerican Samoa (USA) 199 68,688 345.2 Pago Pago, Fagatogo, American SamoaFagatogoFagatogo, American SamoaFagatogo is the seat of government of American Samoa. AS Cook IslandsCook Islands (New ZealandNZ) 240 20,811 86.7 Avarua CK Easter IslandEaster Island (Chile) 163.6 3,791 23.1 Hanga Roa CL French PolynesiaFrench Polynesia (France) 4,167 257,847 61.9 Papeete PF HawaiiHawaii (USA) 16,636 1,283,388 72.8 Honolulu US NiueNiue (NZ) 260 2,134 8.2 Alofi NU Pitcairn IslandsPitcairn Islands (United KingdomUK) 5 47 10 Adamstown, Pitcairn IslandAdamstown PN SamoaSamoa 2,944 179,000 63.2 Apia WS TokelauTokelau (NZ) 10 1,431 143.1 Nukunonu TK TongaTonga 748 106,137 141.9 Nuku'alofaNuku'alofa TO TuvaluTuvalu 26 11,146 428.7 Funafuti TV Wallis and FutunaWallis and Futuna (France) 274 15,585 56.9 Mata-Utu WF Total 8,536,716 35,669,267 4.2 Total minus mainland Australia 849,866 13,641,267 16.1 North PacificNorth Pacific OceanPacific OceanSouth Pacific OceanAustraliaNew ZealandNewZealandChatham IslandsNZHawaiiWake IslandWKFederated States of MicronesiaMicronesiaPalauPapua New GuineaPapua NewGuineaIndonesiaEaster IslandEasterIslandFrench PolynesiaFrenchPolynesiaCook IslandsCookIslandsNew CaledoniaFijiTuvaluKiribatiSolomon IslandsSolomonIslandsTokelauMarshall IslandsMarshallIslandsNauruGalápagos IslandsGalápagosIslandsVanuatuTongaPitcairn IslandsPNGuamNorfolk IslandNFNorthern Mariana IslandsNorthernMarianaSamoaWSAmerican SamoaASWallis and FutunaWFNiueCocos (Keeling) IslandsCCChristmas IslandCXMap of Nations and territories of Oceania incl Australia and New ZealandReligion The predominant religion in Oceania is Christianity. Traditional religions are often animist and prevalent among traditional tribes is the belief in evil spirits (masalai in Tok Pisin), which are blamed for "poisoning" people, causing calamity and death. In recent Australian and New Zealand censuses, large proportions of the population say they belong to "No religion" (which includes humanism, atheism, agnosticism, and rationalism). In Tonga, everyday life is heavily influenced by Polynesian culturePolynesian traditions and especially by the Christian faith. The Bahá'í House of Worship in Tiapapata, Samoa is one of seven designations administered in the Baha'i faith.SportPacific Games The Pacific Games (formerly known as the South Pacific Games) is a multi-sport event, much like the Olympics, (albeit on a much smaller scale), with participation exclusively from countries around the Pacific. It is held every four years and began in 1963. Rugby LeagueRugby league is a popular sport throughout Oceania, and is the national sport of Papua New Guinea "MSN Groups Closure Notice". Groups.msn.com. 2008-10-23. . Retrieved 2009-04-17. (the second most populous country in Oceania after Australia) and is very popular in Australia "Football in Australia - Australia's Culture Portal". Cultureandrecreation.gov.au. 2008-03-28. . Retrieved 2009-04-17. and attracts significant attention across New Zealand and the Pacific Islands. "Rugby League Football - 1966 Encyclopaedia of New Zealand". Teara.govt.nz. 1908-06-13. . Retrieved 2009-04-17.Australia national rugby league teamAustralia and New Zealand national rugby league teamNew Zealand are two of the most successful sides in the world.Wilson, Andy (2009-11-05). "southern hemisphere sides are a class apart". London: guardian.co.uk. . Retrieved 2010-06-17. Australia has won the Rugby League World Cup a record nine times while New Zealand won their first World Cup in 2008 Rugby League World Cup2008. Australia hosted the second tournament in 1957 Rugby League World Cup1957. Australia and New Zealand jointly hosted it in 1968 Rugby League World Cup1968 and 1977 Rugby League World Cup1977. New Zealand hosted the final for the first time in 1985 - 1988 Rugby League World Cup1985 - 1988 tournament and Australia hosted the last tournament in 2008 Rugby League World Cup2008. Rugby UnionFiji playing the Cook Islands at Rugby sevensseven-a-side rugbyRugby union is one of the region's most prominent sports, "Oceania Rugby Vacations". Real Travel. . Retrieved 2009-04-17. and is the national sport of New Zealand, "How many national sports are there". WikiAnswers. . Retrieved 2009-04-17. Samoa, Fiji and Tonga. Fiji national rugby union team (sevens)Fiji's sevens team is one of the most successful in the world, as is New Zealand national rugby union team (sevens)New Zealand's. New Zealand and Australia have won the Rugby World Cup a record two times (tied with South Africa national rugby union teamSouth Africa who have also won it two times). New Zealand won the inaugural World Cup in 1987. Australia and New Zealand jointly hosted the World Cup in 1987. Australia hosted it in 2003 and New Zealand also hosted it in 2011, which they then went on to win. CricketFans' welcome to the Australian team after winning 2007 Cricket World CupCricket is a popular summer sport in Australia and New Zealand. Australia had ruled International cricket as the number one team for more than a decade, and have won four Cricket World Cups and have been runner-up for two times, making them the most successful cricket team. New Zealand is also considered a strong competitor in the sport, with the New Zealand Cricket Team, also called the Black Caps, enjoying success in many competitions. Both Australia and New Zealand are List of International Cricket Council members#Full MembersFull members of the International Cricket CouncilICC. Fiji, Vanuatu and Papua New Guinea are some of the List of International Cricket Council membersAssociate/Affiliate members of the ICC from Oceania that are governed by ICC East Asia-Pacific. Beach Cricket, a greatly simplified variant of cricket played on a sand beach, is also a popular recreational sport in Australia. Cricket is culturally a significant sport for summer in Oceania. The Boxing Day Test is very popular in Australia, conducted every year on 26 December at the Melbourne Cricket Ground, Melbourne. Australian rules footballAustralian rules football is the national sport in Nauru "Nauru AFL team to play in International Cup". solomonstarnews.com. 2008-04-16. . Retrieved 2009-04-17. and is the most popular football code in Australia. "Australian rules football (sport) - Britannica Online Encyclopedia". Britannica.com. . Retrieved 2009-04-17. It is also popular in Papua New Guinea. "pure AFL ... purely Papua New Guinea". Afl Png. . Retrieved 2009-04-17.Association football (soccer) The Oceania Football Confederation (OFC) is one of six association football confederations "FIFA confederations". Fifa.com. . Retrieved 2009-04-17. under the auspices of FIFA, the international governing body of the sport. The OFC is the only confederation without an automatic qualification to the FIFA World CupWorld Cup Finals. Currently the winner of the OFC qualification tournament must play off against an Asian Football ConfederationAsian confederation side to qualify for the World Cup. FIFA world cup

2010 - Oceania preliminary competition "FIFA world cup 2010 - qualifying rounds and places available by confederation". Fifa.com. 2009-04-03. . Retrieved 2009-04-17.Currently, Vanuatu is the only country in Oceania to call football (soccer) its national sport. Oceania has been represented at four World Cup finals tournaments — Australia national association football teamAustralia in 1974 FIFA World Cup1974, 2006 FIFA World Cup2006 and 2010 FIFA World Cup2010, and New Zealand national football teamNew Zealand in 1982 FIFA World Cup1982 and 2010 FIFA World Cup2010. In 2006, Australia joined the Asian Football Confederation and qualified for the 2010 World cup as an Asian entrant. New Zealand qualified through the Oceania Confederation, winning its playoff against Bahrain. 2010 was the first time two countries from Oceania had qualified at the same time, albeit through different conferences. See alsoArt of OceaniaList of cities in OceaniaEconomy of OceaniaMilitary history of OceaniaEuropeans in OceaniaOceania (journal)Festival of Pacific ArtsPacific Islands ForumFlags of OceaniaPacific UnionGeography of OceaniaSecretariat of the Pacific CommunityHistory of OceaniaUnited Nations geoscheme for OceaniaNotesExternal links Oceania at the Open Directory Project "Australia and Oceania" from National Geographickbd:Хы Тlыгуей gag:Okeaniya koi:Океания rue:Океанія nso:Oceania

President_of_Palau

President of the Republic of the Palau	
Incumbent **Johnson Toribiong** since 15 January 2009	
Term length	Four years, renewable once
Inaugural holder	Haruo Remeliik
Formation	2 March 1981

List of Presidents of Palau

No.	Incumbent	Tenure	
		Took Office	Left Office
1	Haruo Remeliik	2 March 1981	30 June 1985
	Thomas Remengesau, Sr. (acting)	30 June 1985	2 July 1985
2	Alfonso Oiterong	2 July 1985	25 October 1985
3	Lazarus Salii	25 October 1985	20 August 1988
	Thomas Remengesau, Sr. (acting)	20 August 1988	1 January 1989
4	Ngiratkel Etpison	1 January 1989	1 January 1993
5	Kuniwo Nakamura	1 January 1993	1 January 2001
6	Thomas Remengesau, Jr.	1 January 2001	15 January 2009
7	Johnson Toribiong	15 January 2009[1]	Present

Latest election

Palauan presidential election, 2008[2] [3]					
Candidate	Running mate	First round		Second round	
		Votes	Percentage	Votes	Percentage
Johnson Toribiong	**Kerai Mariur**	2,526	27.49%	**4,942**	**51.12%**
Elias Chin	Alan Seid	**3,027**	**32.95%**	4,726	48.88%
Surangel Whipps	Billy Kuartei	2,248	24.47%		
Joshua Koshiba	Jackson Ngiraingas	1,387	15.10%		
Invalid or blank votes					
Totals		**9,188**	**100.00%**	**9,668**	**100.00%**
Voter turnout					

See also

- Vice-President of Palau
- High Commissioner of the Trust Territory of the Pacific Islands

References

[1] "Secretary Salazar Meets with Palau President Toribiong March 12, 2009" (http://www.doi.gov/oia/press/2009/03122009.html). *United States Department of the Interior*. 2009-03-12. . Retrieved 2009-04-21.

[2] Miho, David (2008-10-01). "Absentee Ballots Don't Change Outcome Of Palau Presidential Primary" (http://www.pacificmagazine.net/news/2008/10/01/absentee-ballots-dont-change-outcome-of-palau-presidential-primary). *Pacific Magazine*. . Retrieved 2008-10-26.

[3] Miho, David (2008-11-07). "Johnson Toribiong Wins Palau Presidential Race" (http://www.pacificmagazine.net/news/2008/11/07/johnson-toribiong-wins-palau-presidential-race). *Pacific Magazine*. . Retrieved 2008-11-06.

House_of_Delegates

The **House of Delegates** is the name given to the lower house of the legislature in three U.S. states – Maryland, Virginia, and West Virginia.

House of Delegates may also refer to:

- The governing body of several professional organizations, particularly medical ones such as the American Medical Association and its state affiliates is termed "House of Delegates"
- In South Africa under apartheid, the House of Delegates was the House for the country's Indian population in the tricameral Parliament between 1984 and 1994; Amichand Rajbansi was Chief Minister of the House of Delegates tricameral parliamentary chamber
- The House of Delegates is the name of the lower house of the Palau National Congress

Foreign_relations_of_Palau

Palau gained its independence October 1, 1994, with the entry into force of the Compact of Free Association with the United States. Palau was the last Trust Territory of the Pacific Islands territories to gain its independence. Under the Compact, the U.S. remains responsible for Palau's defense for 50 years.

Palau is a sovereign nation and conducts its own foreign relations. Since independence, Palau has established diplomatic relations with a number of nations, including many of its Pacific neighbors. Palau was admitted to the United Nations on December 15, 1994, and has since joined several other international organizations. In 2004, Stuart Beck was appointed to serve as Palau's first permanent representative to the United Nations. Along with the other former Trust Territories, Palau is one of a handful of countries that regularly votes with Israel in the United Nations General Assembly. Palau established relations with the Republic of China on Taiwan, instead of the People's Republic of China, in 1999. Palau established relations with India in April 1995.

The Republic of China (Taiwan), Japan, Philippines, and the United States of America have embassies in Palau.

Palau maintains embassies in the Philippines, Republic of China (Taiwan), Japan, and the USA.

Palau established formal consular links with Israel, it was reported in early 2008.

India

As per the Ministry of External Affairs of India , *India established diplomatic relations with Palau in April 1995. Development* assistance from India has included a grant of US $ 149,841 for purchase of kitchen equipment for Palau National Hospital, US $ in 2008 for purchase of a boat and two pick-up trucks, US$ 100,000 in June 2010 for purchase of computers and grant of 2 ITEC scholarships in 2010-11. Palau has been supportive of issues of importance to India, particularly Indian candidature to international organizations. It supported India's candidature for the non-permanent membership of the UN Security Council for the 2011-12 term. Presently, there are about 15 Indian nationals in Palau.[1]

References

[1] http://meaindia.nic.in/mystart.php?id=500415682

External links

- Republic of Palau Mission to the United Nations (http://www.palauun.org/)

Reprentations of Palau to other nations

- Embassy of Palau in Washington D.C. (http://www.palauembassy.com/home.htm)

Representations of other nations to Palau

- Australian Embassy in Pohnpei (http://www.australianembassy.fm/) Note:Handles relations with Palau

Consular links with Israel:

- http://pidp.eastwestcenter.org/pireport/2008/January/01-04-11.htm

History_of_Palau

Palau was initially settled over 3,000 years ago, and perhaps 4,500 years ago, probably by migrants from the Philippines. British traders became prominent visitors in the 18th century, followed by expanding Spanish influence in the 19th century. Following its defeat in the Spanish-American War, Spain sold Palau and most of the rest of the Caroline Islands to Germany in 1899. Control passed to Japan in 1914 and during World War II the islands were taken by the United States in 1944, with the costly Battle of Peleliu between September 15 and November 25 with more than 2,000 Americans and 10,000 Japanese killed. The islands passed formally to the United States under United Nations auspices in 1947 as part of the Trust Territory of the Pacific Islands.

Four of the Trust Territory districts formed a single federated Micronesian state in 1979, but the districts of Palau and the Marshall Islands declined to participate. Palau, the westernmost cluster of the Caroline Islands, instead opted for independent status in 1978, approved a new constitution and became the Republic of Palau in 1981, and signed a Compact of Free Association with the United States in 1982. After eight referendums and an amendment to the Palauan constitution, the Compact was ratified in 1993 and went into effect on October 1, 1994, marking Palau independent de jure (after Palau was independent de facto since May 25, 1994, when the trusteeship cancelled).

Legislation making Palau an "offshore" financial center was passed by the Senate in 1998. In 2001, Palau passed its first bank regulation and anti-money laundering laws.

Archaeology

Carbon dating of cave burials show a pygmy population, presumably the result of insular dwarfism, from at least 3,000, and perhaps as long as 4,500 years ago until about 900 years ago (1000–2500 BCE until ca. 1100 CE).[1]

Pre-contact

The Palauan language is an outlier among the Austronesian languages, and so does not shed much light on the origins of the modern population. However, there are some indications that it may derive from the Sunda Islands (modern Indonesia).

For thousands of years, Palauans have had a well established matrilineal society, believed to have descended from Javanese precedents. Traditionally land, money, and titles passed through the female line. Clan lands continue to be passed through titled women and first daughters[2] but there is also a modern patrilineal sentiment introduced by imperial Japan. The Japanese government attempted to confiscate and redistribute tribal land into personal ownership during World War II, and there has been little attempt to restore the old order. Legal entanglements continue amongst the various clans.

European contact

Historians take note of the early navigational routes of European explorers in the Pacific. There is disagreement as to whether Spaniard Ruy López de Villalobos, who landed in several Caroline Islands, spotted the Palau archipelago in 1543. No conclusive evidence exists, but some believe he could have seen the tip of a southernmost island in the group.

View of part of the town of Pelew, and the place of Council, 1788

Palau had limited relations before the 18th century, mainly with Yap and Java. Had it not been for shipwrecked islanders who took refuge in the Philippines, Europeans likely would not have found Palau until much later. Englishman Henry Wilson, captain of the East India Company's packet *Antelope*, was shipwrecked off the island of Ulong in 1783. The King of Palau allowed Captain Wilson to take his son, Prince Lee Boo, to England, where he arrived in 1784. However, the prince died soon after of smallpox. The East India Company erected a monument over his grave in St Mary's Churchyard, Rotherhithe. It was Wilson who gave the archipelago the name "Pelew Islands".

Spanish rule

In the late 19th century, possession of the islands was claimed by Britain, Spain, and Imperial Germany. In 1885, the matter was brought to Pope Leo XIII for a decision. The Pope recognized the Spanish claim, but granted economic concessions to Britain and Germany. Palau then became part of the Spanish East Indies, along with the Mariana Islands, the Marshall Islands. and the rest of the Caroline Islands. They were all administered from the Philippines.

Under the Spanish administration, Palau underwent its first forms of extreme foreign influence. Catholicism is the most highlighted amongst all these forms, having been introduced to the native population and accepted rather easily, despite the fact that Palauans were ardent followers of their traditional faith. Elements from the Spanish language were also integrated into the Palauan lexicon; the Palauan word for hammer, "martiliong", is derived from the Spanish "martillo". Even with these forms of influence, Palau has predominantly very little influence from the Spanish occupation compared to other islands within the region, such as Guam. This is partly due to the reason that, aside from Spanish Jesuit missionaries, Spain had very little economic and colonization activity on the islands.

After being defeated in 1898 in the Spanish-American War and losing possession of the Philippine Islands, Spain sold the Palau archipelago to Imperial Germany in the 1899 German-Spanish Treaty.[3] [4]

German era

Palau during German colonization

Koror during the Japanese Mandate

Palau was administered from German New Guinea, and a period of economic development began. German engineers began exploiting the islands' deposits of bauxite and phosphate, and a rich harvest in copra was made.

Although the German occupation had lasted a mere 15 years, major changes in Palauan society had occurred. Many traditional practices among the natives had been outlawed by German administrators, specifically the acts of tattooing and *mengol*. Tattooing had been a practice reserved for members of higher castes, and the Spanish had tried on numerous occasions to banish the practice. Unlike the Spanish, the Germans had a major foothold on the local traditional government, given the economic benefits many clan heads were gaining. The art of tattooing was banished completely by the late 19th century, and the last tattooed native died in the early 1960s.

The practice of *mengol* is one very unique to the islands, and one that was greatly discouraged by all foreign administrations. In *mengol* young unmarried women from one village are sent to male clubhouses in another or other villages, whether enemy or ally, as a sign of peace or friendship. The actions of these *mengol* ranged from simple companionship of the men of the clubhouse, much like the Japanese geisha, to sexual favors. It must be noted however, that the *mengol* were not prostitues, as even the unmarried daughters of clan heads could be sent off as *mengol*.

Japanese mandate

Under the terms of the Anglo-Japanese Alliance, the Empire of Japan declared war on the German Empire in 1914 and invaded German overseas territories in the Pacific Ocean. Palau was seized by ships of the Japanese navy. After the war, the League of Nations awarded Palau to Japan as a Class C League of Nations Mandate.[5]

Japan incorporated the islands as an integral part of its empire, establishing the Nanyo-cho government with Koror Island as the capital.[6] From 1914 to 1922, the Japanese Imperial Navy had been in control. Civilian control was introduced from 1922, and Palau was one of six administrative districts within the Mandate. Japan mounted an aggressive economic development program and promoted large scale immigration by Japanese, Okinawans and Koreans. Native Palauans soon became a small minority in their own homeland. The Japanese continued the German mining activities, and also established bonito (skipjack tuna) canning and copra processing plants in Palau. Japanese economic activity was so extreme during their occupation, that the industrial capital, Koror, was once termed by Japanese officials and tourists as "chiisai Tokyo", meaning "little Tokyo."

The Japanese had perhaps the most profound impact on Palauan society. Under the Japanese government, Palau underwent extreme social, economic, and political changes. Unlike other parts of the Japanese Empire, the local Japanese administration recognized the benefits of utilizing the traditional government to their advantage, given the extreme adherence the natives followed towards their clans and traditions. Under the Japanese government, heads of the Palauan traditional government were replaced by more "Japanese" natives, as a way of securing the obedience of the Palauan people. This proved rather successful, as more and more Palauans so the benefits the Japanese government was providing. Under the Japanese administration, all Palauan children were required to attend school, and by the 1930s, nearly all Palauan children were literate in Japanese, with the exception of children in outlier islands, specifically the Southwest Islands.

World War II

The Japanese presence made Palau a major target for the Allied forces in World War II. Peleliu was a scene of intense fighting between American and Japanese forces in 1944. The battle ended in an Allied victory, but at a high cost for both sides. All surviving Japanese were repatriated after the end of the war. There are still about 100 American servicemen listed as Missing In Action in Palau. Starting in 1993, a small group of American volunteers called The BentProp Project has searched the waters and jungles of Palau for information that could lead to the identification and recovery of these remains.

Post-war development

In 1947, the United Nations decided the United States would administer Palau as part of the Trust Territory of the Pacific Islands. In 1979, Palauans voted against joining the Federated States of Micronesia because of language and cultural differences. A long period of transition occurred between 1979 and 1994, specifically in the capital, Koror, on the topic of relations with the United States. During this period, Palau underwent its first form of civil unrest since the last inter-village war in 1879. This civil unrest had escalated into extreme violence, resulting in numerous bombings of homes of powerful politicians, intellectuals, influential locals, and the death of two presidents. At one point, the civil struggle had resulted to the point of anarchy, and the King of the Southern Federation of Palau, the Ibedul, assumed responsibility over governing the Palauan islands which thereby established a temporary state of absolute monarchy. After a long period of transition, including the violent deaths of two presidents (Haruo Remeliik in 1985 and Lazarus Salii in 1988), Palau voted in 1994 to freely associate with the United States while retaining independence under the Compact of Free Association.

References

[1] Lee R. Berger, Steven E. Churchill, Bonita De Klerk1, Rhonda L. Quinn (March 2008). "Small-Bodied Humans from Palau, Micronesia". *PLoS ONE* **3** (3): e1780. doi:10.1371/journal.pone.0001780. PMC 2268239. PMID 18347737.

[2] Palau National Communications Corporation (http://www.palaunet.com/pw_culture.aspx)

[3] Sandafayre.com on Palauan history (http://www.sandafayre.com/atlas/carois.htm)

[4] United States Department of State article on Palau (http://www.state.gov/r/pa/ei/bgn/1840.htm)

[5] Peatty, Nan'Yo: The Rise and Fall of the Japanese in Micronesia

[6] Beasley, Japanese Imperialism

The New Capitol in Palau

External links

- U.S. State Department Background Note: Palau (http://www.state.gov/r/pa/ei/bgn/1840.htm)

Economy_of_Palau

The **economy of Palau** consists primarily of subsistence agriculture and fishing. The government is the major employer of the work force, relying heavily on financial assistance from the United States. The population enjoys a per capita income of more than twice that of the Philippines and much of Micronesia. Long-term prospects for the tourist sector have been greatly bolstered by the expansion of air travel in the Pacific and the rising prosperity of leading East Asian countries.

Rock Islands in Palau - Such scenery has helped invite tourism and build up Palau's economy.

Palau's per capita GDP of $8,900 makes it one of the wealthier Pacific Island states. Nominal GDP increased by an annual average of nearly 14% from 1983 to 1990, and by an annual rate of over 10% from 1991 to 1997. Growth turned sharply negative in 1998 and 1999 as a result of the 1997 Asian financial crisis.

Industries

Tourism is Palau's main industry. Activity focuses on scuba diving and snorkeling among the islands' rich marine environment, including the Floating Garden Islands to the west of Koror. The number of visitors—85% of whom come from Japan, Taiwan, and the U.S.—reached nearly 67,000 in 1997, more than quadruple the level of a decade earlier. Tourism earned $67 million in foreign exchange for Palau in 1996, accounting for roughly half of GDP. Arrivals from Asian countries dropped in 1998 and 1999 due to the regional economic downturn and the depreciation of many Asian currencies against the dollar, which made Palau's dollar-denominated prices more expensive.

The service sector dominates the Palauan economy, contributing more than 80% of GDP and employing three-quarters of the work force. The government alone employs nearly 30% of workers. One of the government's main responsibilities is administering external assistance. Under the terms of the Compact of Free Association with the United States, Palau will receive more than $450 million in assistance over 15 years and is eligible to participate in more than 40 federal programs. The first grant of $142 million was made in 1994. Further annual payments in lesser amounts will be made through 2009. U.S. grants in 1999 totaled $24 million.

Construction is the most important industrial activity, contributing over 9% of GDP. Several large infrastructure projects, including the rebuilding of the bridge connecting Koror and Babeldaob Islands after its collapse in 1996 and the construction of a highway around the rim of Babeldaob, boosted activity at the end of 1990s.

Agriculture is mainly on a subsistence level, the principal crops being coconuts, root crops, and bananas. Fishing is a potential source of revenue, but the islands' tuna output dropped by over one-third during the 1990s.

There are no Patent Laws in Palau. [1]

Economic challenges

The main economic challenge confronting Palau is to ensure the long-term viability of its economy by reducing its reliance on foreign assistance. Palau has created a trust fund to be drawn upon after the cessation of Compact grants, the value of which had grown to $140 million by the beginning of 2009.

Economic statistics

Income

Gross Domestic Product (GDP): purchasing power parity - $132 million (2009 est.) (GDP estimate includes U.S. subsidy)

GDP - real growth rate: 1% (2009 est.)

GDP - per capita: purchasing power parity - $8 500 (2009 est.)

GDP - composition by sector

- *agriculture*: N/A
- *industry*: N/A
- *services*: N/A

Population below poverty line: N/A

Household income or consumption by percentage share

- *lowest 10%*: N/A
- *highest 10%*: N/A

Inflation

Inflation rate (consumer prices): 3.4% (2000 est.)

Workforce

Labor force: 10 200 (2000)

Labor force - by occupation

- *Agriculture*: 20%
- *Industry*: N/A
- *Services*: N/A

(1990)

Unemployment rate: 4.2% (2000 est.)

Budget

- *Revenues*: $57.7 million
- *Expenditures*: $80.8 million, including capital expenditures of $17.1 million (FY 98/99 est.)

Industries

Dominant industries include tourism, craft items (from shell, wood, pearls), construction, and garment making.

Industrial production growth rate: N/A

Electricity

Electricity - production: 200 MWh (1996)

Electricity - production by source

- *Fossil fuels*: 85%
- *Hydro*: 15%
- *Nuclear*: 0%
- *Other*: 0%

(1996)

Electricity - consumption: 200 MWh (1996)

Electricity - exports: 0 kWh (1996)

Electricity - imports: 0 kWh (1996)

Trade

Agriculture - products: coconuts, copra, cassava (tapioca), sweet potatoes

Exports: $18 million (f.o.b., 2001)

Exports - commodities: trochus (type of shellfish), tuna, copra, handicrafts

Exports - partners: U.S., Japan, Singapore (2000)

Imports: $99 million (f.o.b., 2001 est.)

Imports - commodities: machinery and equipment, fuels, metals, foodstuffs

Imports - partners: U.S., Guam, Japan, Singapore, Korea (2000)

Debt and aid

Debt - external: $57 million (FY08/09)

Economic aid - recipient: $155.8 million. Note: the Compact of Free Association with the U.S., entered into after the end of the UN trusteeship on 1 October 1994, provides Palau with up to $700 million in U.S. aid over 15 years in return for furnishing military facilities.

Currency

1 United States dollar (US$) = 100 cents

Exchange rates: U.S. currency is used

Fiscal year

1 October – 30 September

See also

- Palau

Article Sources and Contributors

Outline_of_Palau *Source*: http://en.wikipedia.org/w/index.php?title=Outline_of_Palau *Contributors*: AK Auto, Alan Liefting, Buaidh, ChiragPatnaik, CommonsDelinker, Danger, Funandtrvl, Highfields, Juliancolton, Kintetsubuffalo, Malinaccier, Matthewedwards, Plastikspork, Rich Farmbrough, Svick, Tarheel95, The Transhumanist, Thehelpfulone, 1 anonymous edits

United_Nations_Trusteeship_Council *Source*: http://en.wikipedia.org/w/index.php?title=United_Nations_Trusteeship_Council *Contributors*: Alanmak, AnnaFrance, Arjayay, BTLizard, Beland, Bergsten, Bobblewik, BomBom, Coemgenus, Colonies Chris, Conversion script, Darker Dreams, Deror avi, Dpmuk, Ed g2s, Efe, El C, Empoor, Erpbridge, EvilSuggestions, Flambelle, Formulax, Good Olfactory, Gryffindor, Gsandi, Instantnood, J.reed, JHunterJ, Jezarnold, John K, Joseph Solis in Australia, Jusjih, Karen Johnson, Kurykh, Lapicero, Lightmouse, Mcg21, Mclay1, Mic, Narutolovehinata5, Newyorkbrad, Omdo, PaulHanson, Postdlf, Pras, Raafael, Rad Racer, Raul654, Rayshix, Rd232, Rich Farmbrough, Richard75, Rodsan18, Romanm, SJK, Saebvn, Sergeant Cribb, Seth Ilys, Sl, That-Vela-Fella, The Evil IP address, Tipiac, Tom harrison, TreasuryTag, Милан Јелисавчић, 53 anonymous edits

Philippines *Source*: http://en.wikipedia.org/w/index.php?title=Philippines *Contributors*: $antander, (jarbarf), *drew, 0oflasho0, 100110100, 1234596asdfgh, 123eWick, 138.89.101.xxx, 16@r, 23prootie, 24fan24, 2D, 2d0k, 3.217, 334a, 55david, 9shaun, A2Kafir, A3r0, ABF, ACSE, AMP408, ANNNGUYEN, ASDFGH, Aaker, Aaronp808, Abfall-Reiniger, About the Sultan, Acdeleon, Aclarado, Acolombo1, Addicted04, Addshore, Adelina and Hannah, Adinemb, Adraeus, Adrian, Adrianvill, Adzuband, Aeron Valderrama, Ahoerstemeier, Ahuskay, Airconswitch, Aitias, Akanemoto, AkoDanielle, Akuindo, Al-Andalus, Alan Liefting, Alansohn, Albert584, Albertc1990, Aleenf1, AlefZet, Alex.tan, Alex43223, Alexanderj, Alexius08, AlexiusHoratius, Allentchang, Allissonn, Allstarecho, AlvarezQz, Amante de la Paz, Amicuspublilius, Ams80, Amy 5580, Amzimuthis, Anak 1, Anarchangel, Ancheta Wis, Andersonbd1, Andonic, Andrew0921, Andrewlp1991, Andy Marchbanks, Anetode, Angela, Angelo6397, Anger22, Anietor, Ankit jn, Ann Stouter, Anon39923, Anonymous Dissident, Anonymous editor, Antandrus, Antarcticaaacold, Anthony Appleyard, Antiedman, Antipoeten, AntonioMartin, AnyPerson, Anyo Niminus, Aph1ruderman, Apollo1758, Arakunem, Arathjp, Arbitrarily0, Archtransit, Ardonik, ArglebargleIV, Ariellim, Arjayay, ArmadilloFromHell, Arnoutf, Art LaPella, Ashleyisachild, Ashmoo, Ashwinr, Asterion, Astuishin, AtD, AtaruMoroboshi, Atlasquery, Atomicarrow, Attic Owl, Aude, Audiosmurf, Auroro Del Rosario, Australia boy, Australian Jezza, Australian Matt, Avala, Avnjay, Awgibbons, AxelBoldt, Axxand, Az1568, Azndragon126, Azurich, BGordon, BSATwinTowers, Baa, Bacierra, Backslash Forwardslash, Bacolod Global, BadBlueBoy, Badagnani, Badboingboy, Baeksu, Banaticus, Bantilan, BanyanTree, Barron52, Barryob, Basser g, Bazonka, Bdviets, Beano, Beijingolympics, Bemasc, Ben Ben, BenGriffiths, Bennjohncatipay, Berserkerz Crit, Betacommand, Bhadani, Bickeringwife, Billsta1, Biruitorul, Bitefusion, Bjmarfito, Black Falcon, Blackfinale, Blackspotw, Blood Red Sandman, Bloodtm, Bluelip, Bluemask, Bmdavll, Bobblewik, Bobguy7, Bobo192, Bobong pinoy, Bogdan, Bogey97, Bonadea, Bone1234, Bongwarrior, Bonzon2, Bookandcoffee, Boud, Boxedor, Brian0918, BrianZhukov, Brianboulton, Brianmagallano, Brianski, BritishWatcher, Brownpau, Bruce Hall, Btnheazy03, Buaidh, Bucephalus, Buckshot06, Buhay Tao, Bulilit, Bull Market, Buonaparte69, Burndownthedisco, Buszmail, Buwayahman, C.Fred, C.Kent87, C777, CJLL Wright, CJSpooks, COOL JHX, CST, CWii, Caiaffa, Cajunfusion, Cal 1234, CalJW, Cali567, CalicoCatLover, Caltas, CambridgeBayWeather, Camembert, Camtay123, Can't sleep, clown will eat me, CanadianLinuxUser, Caniago, CanisRufus, Cantus, Caponer, Capricorn42, Capriugm, Carl9331, Carlo ms06, Carlo9496, CashRules, Catdude, Catgut, Causa sui, Cbdorsett, Cbesancon, Ccirulli, Cdcjr, Cebuexpat, Cenwin88lee, Chan Han Xiang, CharlieSierra, CharlotteWebb, Chasingsol, Chattamaca, CheMechanical, Ched Davis, Cheddlemsic, Cheezypants, Chensiyuan, Chenzw, Chiavenir, Chibler, Chicago god, Chicbicyclist, Chick Bowen, Chiklit, Chilledsunshine, Chinchanchon, Chipmunkdavis, Chocoforfriends, Chodorkovskiy, Chowells, Chris the speller, Chrishmt0423, Christian Mac Juane, Christmasfairy, Christopher Pritchard, Christopher Sundita, Chriswiki, CiTrusD, Ciaomarco, CieloEstrellado, Cindytrinidad, Circa 1900, Circeus, Citicat, Cjh, ClanCC, Clawson, Closedmouth, Cloud0211, Cma, Cmichael, Cohesion, Colonies Chris, CombatCraig, Cometstyles, CommonsDelinker, Compgeo.98, Connormah, Conte di Cavour, Contributor14, Conversion script, Coojah, Coolwhiz, Corecirculator, Corusant, Coupdeforce, Cracker017, Cratbro, CrazieXninja, Credema, Cremepuff222, Critic-at-Arms, Cromwellt, Cs-wolves, Cst17, Cuak's, Cubao-Filipinas, Cuchullain, Curps, Cyberpaul, DCrhD, DJFrankie2468, DMacks, DReyn86, DabMachine, Dabomb87, Dabryanman, Daduzi, Daguhoy, Daimengrui, Dakilang Isagani, Dangme, Daniel5127, DantePh, Dar book, Dark Tea, Darp-a-parp, Darrentzw, Darth Narutorious, Darth Panda, Dataleft, Datu, Dave6, Daverocks, David G Brault, David Kernow, Davidcannon, Davinciscode, Dawfromdaw, Dawnseeker2000, Dcmcgov, Ddstretch, DeadEyeArrow, Debresser, Deepsweech, Deflective, Dekimasu, Dekisugi, Delldot, Deor, DerHexer, Derek.cashman, Deviathan, Devil Master, Dick Kimball, Digthis, Dina, Dineshmohnani, Dinocarlo, Discospinster, Distal24, DivineIntervention, Diwas, Djinn112, Djln, Dlohcierekim's sock, Dmaestoso, DmsanjuanMM3, Doc glasgow, Doc9871, Docu, Doggio!, Dominus, Donald Albury, DondiMapa, Donnpascual, DopefishJustin, Doublet89, Doulos Christos, Dove1950, Dpv, Dragonbeast, DragosteaDinTei, Drbug, Drilnoth, Dxmillenium, Dysepsion, E Wing, EJF, ESkog, EVula, Earl.phil, Earth, Earth88888, East718, Ebizur, Ed, Edbon3000, Editor232, Editore99, Edivorce, Edmundolee, Edpudol, Efe, Ehn, Eivind F Øyangen, Ejrrjs, Ekim sevilla, Ekong13, El C, El Caudillo, El aprendelenguas, El dia, Electionworld, Electromagnetictop, Elektrik blue 82, Elf ideas, Elm, Elockid, ElockidAlternate, Emjoven, Emperor Genius, Emperork, Emx, EncMstr, Enchanter, EncycloPetey, Encyclopedist, Enigmaman, Enock41, Enviroboy, Enzo Aquarius, Enzoamp, Enzomartinelli, EoGuy, Epbr123, Epidemic, Epolk, Epzy, Eric Wester, Erikhants, Error slow, Escape Orbit, EscapingLife, Esimal, Esoglou, Esrever, Eternal dragon, Euchiasmus, Eugene van der Pijll, Eurogirl3, Euronbaja, Evan fornilda, EvanProdromou, Everyking, Evil Monkey, Evil wombat, Excellent 15, Excellent15, Excirial, Exec8, Explicit, Ezhiki, Fabulous Creature, Fabyan17, FadulJoseA, Faithlessthewonderboy, Falconleaf, Faradayplank, FayssalF, Felipito1.966, Fenice, Ferio252, FiLoCo, Fibonacci, Fiet Nam, Fifteencounts, Fighter 10, Filelakeshoe, Filipinayzd, FionaBelinda, Fireice, Fireineyes, Firstdivider, Fishnet37222, Flatterworld, Flewis, Flip89, Florancio, Florentino floro, Flowerparty, Folksong, ForgottenManC, Fr3shmen, Fram, FrancoGG, Freakofnurture, FreeMorpheme, FreplySpang, Freska123, Friginator, Fummyiu, Fungusperson, Funnyfarmofdoom, Furball, Fuzzzone, Fvw, G Rose, G. Campbell, GHe, Gadfium, Gaff, Gaimhreadhan, Gaius Cornelius, Galaxiaad, Galwhaa, Gamer007, Gardar Rurak, Gareon, Gargolla, Garion96, Garzo, Gaurav1146, Gaz, Gdavidp1, Geemics, GelloX, Gelo17, Genericuser69, Geoff Plourde, GeorgeOrr, Gerald Farinas, Gggh, Giant89, Gidonb, Gilgamesh, Gilliam, Gilligan Meriman, Gimferrer, Gintong Liwanag Ng Araw, Giuliani2008, Glacier Wolf, Glane23, GlaucusAtlanticus, Glenncando, Glenthadriel123, Glimz, Gloomy gal, Gogo Dodo, GoingBatty, Golbez, Goldfishbutt, Gonzo fan2007, Good Olfactory, GoodDay, GoonerDP, Got Milked, Gr8opinionater, GraemeL, Grafen, Graham87, Grant.harbison, Grapple X, Green Giant, Greenguy1090, Gregory Watson, Grenavitar, Grendelkhan, Gronky, Gruentag, Grunt, Gryffindor, Gsarwa, Gtrmp, Gubernatoria, Guillom, Gulusmarzo@yahoo.com, Guluspogi@yahoo.com, Gundamforce, Guptadeepak, Gurch, Gurubrahma, Gwalla, Gwernol, Gzornenplatz, Gökhan, HJ Mitchell, HZYrd, Haakon, Hadal, HakkiOzbasli, Hakz2007, Haldful, Handsome xxx, Handtell, Hantzen, HappyCamper, Harfarhs, HariezNor, Harlock jds, Harryboyles, Harryzilber, Hawkestone, Hayden5650, Hdt83, Heartbreaker21, Heartthrob god, HeikoEvermann, Heimstern, Heisenbergthechemist, HeleneSylvie, Hellomota, Henry Flower, Henry01, HenryLi, Hensu75, HexaChord, Highfields, Highvale, Hihellowhatsup, Hinicome, Hintha, Hippo43, Hmains, Homagetocatalonia, Hongthay, Hotcrocodile, Howard the Duck, Hu, Huntster, Husond, Hut 8.5, Hydrogen Iodide, Hyperpain2000, I Love You!!!™, I c linguist, Iamjameson, Iamnot, Iamtheironman, Ian77, Icemanofbarcelona101, Icon7, Idont Havaname, Ief, IeieieieFrenchenenenene, Igiffin, Ikabod08, Ike74ph, Ilampour, Ilion2, Improv, Inaeldi, Infrogmation, Innapoy, Insanephantom, Insanity Incarnate, Instinct, Into The Fray, Introductory adverb clause, Invertzoo, Iridescent, Irrypride, Isko1901, Iwanttoeditthissh, Ixfd64, J.delanoy, JAY3LLE, JCRB, JEREMYBB, JForget, JL 09, JLaTondre, JNW, JOEFIXIT, JParis, JZTine, Jac16888, Jacek Kendysz, Jack Phoenix, Jacksav, Jacopone, Jademoonrose, JadziaLover, Jagged 85, Jake Wartenberg, Jakezing, Jaldgreat, JamesR, Jamesam16, Jamesdjcr15, Jameswilson, Jan eissfeldt, Jandolk, Janjoe, Janster15, Japee, Jaranda, Javierbaires1, Jayjay042298456, Jayv deut, Jbvillarante, Jc Altura, Jccalhoun, Jchan06, Jchitto, Jcs1990, Jdcjds, Jeff G., Jeffrey Batac, Jello8, Jeltz, Jenmoa, Jennna, Jep1793, Jepoy junkie, Jeppesn, Jessellao207, Jezarnold, Jhendin, Jhf, Jiang, Jibco, Jim10701, Jizzy p, Jj137, Jjhcap99, Jma71, Jmaddux23, Jni, JoJan, JoSePh, Joachim Weckermann, JoanneB, Joao Xavier, Joe9320, Joeabraham2005, John Hill, JohnDR, JohnI, Johnbrillantes, Johndarrington, Johndelapena, Johnlemartirao, Johnnie ong, Johnny2035, JohnnyTwain, JohnsonSia, Jojit fb, Jolaspc, Jolo887, Jonarge, JonasEncyclopedia, Jondel, Jonel002, Jonica c, Jorgecortz122, Jorunn, Joseph Solis in Australia, Josh Parris, Joshdboz, Joshua rivas, Jossi, Joyous!, Jpfards, Jpgarciamd, Jpgordon, Jpittman, Jpople, Jpramosrn, Jpuligan 12, Jpvinall, Jrboi, Jrboi96786, Jrdioko, Jredmond, Jtalledo, JuJube, Juansuing, Juliancolton, Junamellabon, Just Another Dan, Jute whatever, Jza84, K3k 04 16, KC Panchal, KaElin, Kablammo, Kagoikunai, Kahusi, Kaisershatner, Kaj2134, Karimarie, Karl Meier, Kata Alreshim, Kathanar, KathrynLybarger, Kauffner, Kazikame, Kdpsuarez, Keilana, Kelbert, Kelly Martin, Kelner, Kennethduncan, Kesla, Ketty109, Keyaydikeyeychako, Kguirnela, Khamm, Khoikhoi, Khurg100, Kid45laundry, Kiddie Techie, Kiel friendly, Kiensvay, Kieron, Kimxiao, King Lopez, Kingpin13, Kintetsubuffalo, Kiozuki, Kittenheel, Kjbaldio, Kneepole, Knutux, KobeWanKenobi, Kokoyvp, Kolins, Konsulromanum, Kookykman, Korina iluvu, Kotniski, Koyaanis Qatsi, KoyaanisQatsi, Kozuch, Kpjas, Kraftlos, Kralizec!, Krellis, Kribbeh, Krich, Krizpo, Kukini, Kungfuadam, Kuru, Kusma, Kwamikagami, Kwp0415, Kyh87, L0s3r, L33th4x0rguy, LFIntalan, LWF, La Fuente, La Pianista, La goutte de pluie, Lagalag, Lambanog, Langara College, Larry Sanger, Lassirena6, Latka, Lavdiaz, Ld100, Leapmark, Lejman, Lemchesvej, Lenticel, LeoDaVinci, Leonard^Bloom, Lesanges, Leszek Jańczuk, Levineps, Lexicon, Liberalista, Liezl0724, LightAj, Lightmouse, Lights, Lightspeedchick, Ligulem, Lil2k5kidd, LilHelpa, Lilac Soul, Lilpinoy 82, LinaMishima, Lindmere, Ling.Nut, LinuxPro, Lithpiperpilot, Liza sakura animegirl, Llort, LlywelynII, Lnesa, LonelyMarble, Look2See1, LordAntagonist, Lotje, Lovebus, LovesMacs, Lowzeewee, Lt. penguin, Lucy Taka, Lucyin, Luigiacruz, Luis1972, Luisztdt, Luk, Luna Santin, Lupin, MBK004, MECU, MER-C, MJCdetroit, MKoltnow, MPF, MSGJ, MZMcBride, Mac Davis, Macheccruz, Mackant1, Mackoy20, Magalhães, Magiciandude, Magister Mathematicae, Magsmon, Mailer diablo, Majestic27, MajorPlayer, Majuro, Makemi, Maksim L., Malcolma, Malhonen, Malo, Man vyi, Mang kiko, Mangacha, Marcintexas, Marianne aseo76, Marimik, Mario limos, Mark Ryan, Markxvi, MarsRover, Marskell, MartinJoh00, Martindo, Martinwilke1980, Marzoregulus@yahoo.co.ph, Master Annihilator, Master of Puppets, Masterjamie, Matababoy, Matdem, Materialscientist, MattTM, Matthew Yeager, Matthew06 87, Matthewprc, MattieTK, Mav, Maxamegalon2000, Maxim, Maximus Rex, Maxis ftw, Maxpowers12345, Maxschmelling, Mayanser, Mboverload, McDogm, Mdf, Mean as custard, Mediacenterentertainment, Megumi888, Mehmet cetin100, Melchoir, Melvin87, Memchii, Mendaliv, Meow, Merbabu, Merlin002, Merlot70, Mermaid from the Baltic Sea, Merovingian, Mervynbunique, Mesgul82, Metusdominis, Meursault2004, Mfa fariz, Mia0514, Miaow Miaow, Michaelbarnes, Michaelbueza, Mickey gfss2007, Micro101, Midnightowl, Mightymights, Migu16, Mikael07, Mike211423, Mikeabundo, Mikenassau, Mikhael07, Mikhailderivera, Mikko Paananen, MilborneOne, Miloais, Mingovia, MinnesotanConfederacy, Miramo, Mistakefinder, Mk32, Mm40, Mnomigol, Modda chill, Modemac, Modulatum, Moeron, Mogultown, Moment Deuterium, Moncrief, Monzonda, Morganfitzp, Mori33, Moriori, Mortai, Morwen, Mr. Blue, MrOllie, Mrbellamguel, Mrdarrett, Mricon, Msidn, Mufka, Mugiwaranorufi, Mumble45, Murano333, Museslave, Mushroom, Mwilso24, Mxn, Myanw, Mygerardromance, Mübarek-Demokrasi, NHJG, NJA, NWOG, Naelphin, Najyar, Nakakapagpabagabag, Naniwako, Nash16, Nasugbu Batangas, Nasugbu batangas, Natalie Erin, Nathan92295, Natsirtj, Naturalorganic65, Nautical Mongoose, NawlinWiki, Naxju, Nburden, Neelix, Nescalona, Netoholic, Neurocod, Neutr8, Neutrality, Neverquick, NewEnglandYankee, Newarkmya, NickBush24, Nicke L, Nickshanks, NicolaM, Nicolas1981, Niduzzi, NielsGabriel, Nightstallion, Nihiltres, Nilfanion, Nino Gonzales, Nixeagle, Nkcs, Nlg turbo, Nneonneo, Noah Salzman, Noarmyinphili, Noctibus, Noimnotokay, Nonnifrickatzovindaloohosen, Nono64, Northamerica1000, Northbeach23, Northoltjohn, Notheruser, Notthe600, Novem Linguae, Noypi380, Nposs, Nquotes, Nsaa, Nstott, Nubius, Nucleusboy, Numberoneconti, Numbo3, Nutiketaiel, Nv8200p, Nyttend, Ocatecir, Oda Mari, Of, Ohnoitsjamie, Okeencosc, OldakQuill, Oleg Alexandrov, OliverTwisted, Olivier, Omicronpersei8, Omidkhani, Oni Lukos, Opelio, Optakeover, Opus88888, Orangutan, Orceuos, Oren neu dag, Orioane, Orsahnses, Othanwiki2009, Otisjimmy1, Overrayd, Ovibos, OwenX, Ownedyouux3, Oxfordresearches, Oxymoron83, PDH, PLM Collegian, PLM Community, PTSE, PacoManila, Paine Ellsworth, Pairadox, Pakingshet, Palaris, Paleorthid, Pam519, Pandrio Androtti, Pare Mo, Parkwells, Parthian Scribe, Pastricide, Patstuart, Paul August, Pavone, Paxnominus, Pepe alas, Peregrine981, Peripitus, PeterOram, PeterisP, Pgan002, Pharm unoc, Phichanad, Phil Boswell,

PhilKnight, Philip Trueman, PhilippInfo, Philippinepresidency, Phillip J, Philoyster, Philtro, PhnomPencil, Piano non troppo, PikDig, Pikiwedia, Pilipinoman, Pilot expert, Pilotguy, PinayFLAVALuSHuZ, PinchasC, Pink!Teen, Pinkblue, Pinoiboi7540, Pinoy2298, Pinoy916, PinoyFilAmPride, Pinoyamerican, Pinoyboy, Pip2andahalf, Pjred, Plastikfakkincok, Pne87, PoccilScript, Pointer1, Polaron, Poptart65, Potatoswatter, Powella, Prara, Prats13, Pravdaverita, President Rhapsody, PrestonH, Primalchaos, Prince denison, Prince edward154, Prodego, ProfessorGreg, Profoss, Psyklax, Pupitar13, Pwhitwor, Pyrospirit, Pyrotec, Qasaqsuyu, Qccirclejunkie, Qtoktok, Quadratus, Quebec99, Quess, Quintote, Quuxplusone, Qwanqwa, QzNhc, R Lowry, R000t, R8sam&meg, RJN, RNAlonto, Rabasolo, RacitsPhil, RadJad56, RadiantRay, RadioKirk, Rafael 112683, Rafaelvicentemarquez, RaffyJezrylle, Raigeiki55, RainbowOfLight, Ram-Man, Ramirez72, Rammaum, Ramogiao, RandomP, Rarelibra, Rastapopoulos, Ravens326, Raymundchua, RazorICE, Rbraunwa, Rdsmith4, RebeccaFilardo, RedHillian, RedRollerskate, Redishflag, Rednblu, Reduxtion, RegPSN, Regulus marzo4103, Regulus marzo4103@yahoo.com, Renanschneider, Researcher1982, ResurgamII, Rettetast, Reviewgirlerika, RexNL, Reywas92, Rfcesrce, Rich Farmbrough, RichMac, Richard Relucio, Richard88Bacs, Richfife, Rick Block, RickK, Rikkuwolf, RingtailedFox, Risercher, Riyehn, Rizalninoynapoleon, Rjwilmsi, Rkt2312, Rlquall, Rmcsamson, Rmky87, Rmsharpe, Rob Hooft, Rob Lindsey, Rob122, Robbiemuffin, Robbinato, Robert talan, Robertvan1, Robomaeyhem, Robz1234, Rock4arolla, Roewuck, Rolly2005, Romanm, Rosadobigd, Rose Garden, Rosspz, Rosvenir, Rotational, Roxther, RoyBoy, Royalguard11, Rrdacaychua, Rror, Rtyq2, Ruffyuppie, Ruick25, Rumping, Rushadthomas, Ruslik0, RxS, Ryan nicolas57, RyanCross, Ryomaandres, Ryuharu03, Ryulong, SFC9394, SMcCandlish, SORIANO FAMILY, SUpErPoWeRgIrL, SWAdair, Saforrest, Saikee, Saimdusan, SamEV, Samantha555, Sambuddy, San Fernando Valley, SanCarlosFilipinas, Sandstein, Sandstig, Sannse, Sanya3, Saperaud, Sarcelles, Sardanaphalus, Sardonyx, SarekOfVulcan, Saynoe, Scetoaux, Schlacks, Scipius, Sciurinæ, Scorpion prinz, Scott McNay, Scott Ritchie, Scratch1990, Screennamestaken, Seabass341, Seansabado, Seav, Seb az86556, Sebjarod, Secaundis, Secfan, Secretadmirer 1257, Sejamax, Semmler, Senator Palpatine, Senior Fellow, Seppepogi, Seraphim, Severo, Sg2kca, Shadow1, Shalom Yechiel, Shanes, Sherwin ph, ShlowBABIEE, Shoeofdeath, ShraZr, Shrumster, Shshshsh, Shushruth, Siddhant, Silentaria, Silverxxx, SimonP, Simplifier08, Sir Edgar, Sir Nicholas de Mimsy-Porpington, Sirrodz, Sjakkalle, Sjharten, Sk8ajoe, Skater, Skullcom2010, Sky Harbor, SkyWalker, Skyduster, Slakr, Sleepzyflip, Slowking Man, Smellsofbikes, Smutty71, Snowdog, Soccerissweet995, Softnayr, Soliloquial, Somsinee, Sooner Dave, Soosed, Sopholatre, Soup, Soxwon, Space Cadet, Spangineer, Spcleddy, SpeakFree, Speer320, Spencer, Splash, Spliffy, SpookyMulder, Spry895, SpuriousQ, Sputian 13, Sputnikcccp, Spylab, Squash, SraQJN, Srinivas, StG1990, Stalmannen, Stan Shebs, Stardust8212, Starionwolf, Starwars664, Staxringold, Steakknife96, Steel, Steinbach, Stephen Bain, Stephenb, Stephenginns, Steve Dufour, Stevenphil, Stooppy, Str1977, Street Mix Live Entertainment, Stui, SunKing, Suntag, Super-Magician, Suruena, Svdp, Sven Eriksson, SwedishZeta, SweetCarmen, Synchronism, SyntaxError55, Sysy, THEN WHO WAS PHONE?, TUF-KAT, Tagishsimon, Tainter, Taiwai94, Tam3rd, Tanglewood4, TantalumTelluride, Taospark, Tapir Terrific, Tarong, Tarquin, Tasc, TastyPoutine, Tatarian, Tawker, Tbhotch, Tcncv, TeaDrinker, Teddy8911, Tellyaddict, Template namespace initialisation script, Tennis expert, Terence, Tetracube, Teumo, TeunSpaans, Texture, Thailander, The Haunted Angel, The Lake Effect, The Person Who Is Strange, The Placebo Effect, The REAL Dyslexic Q-Thief, The Thing That Should Not Be, The Way, The birch tree and the dandelion, The ed17, The marman, The sunder king, The undertow, The wub, TheBusiness, TheCoffee, TheCustomOfLife, TheKMan, TheMoth&TheButterfly, TheOtherJesse, TheProject, TheTechieGeek63, Thebeginning, Thecurran, Thegreatdr, Thehelpfulone, Therefore, Therequiembellishere, Theresa knott, Thetriwizardchamp, Thingg, Thinking of England, Thorwald, Tiddly Tom, Tide rolls, Tiger888, Tigershrike, Tim Q. Wells, TimR, Timberer, Timir2, Timlight, Timo Honkasalo, Tinpinkprincess, Tinton5, Tiptoety, Tirid Tirid, Titopao, Titoxd, Tj brewed, Tobby72, Tobias Conradi, Toddst1, Tohd8BohaithuGh1, Tohru Honda13, Toi, Tom Radulovich, Tom harrison, Tomas De Aquino, Tombseye, Tony Sidaway, Tony1, TopQuark, Topbanana, Tpbradbury, Tradnor, Transity, Travelbird, Travis28, TriniMuñoz, Truflip99, Tslocum, Tuminof, Twinxor, USTerritory, Uaghm8, Uchrgseous, Ugen64, Ulric1313, Ungvichian, Unschool, Unyoyega, Updatehelper, User0000, Usergreatpower, Username12, Utcursch, Uthanc, VX, Vaanashe12, Valistar, Van helsing, Vancouverjersey, Vanished User 1004, Vaquero100, Vardion, VegaDark, Vegaswikian, Velho, Ventada de Manila, Vera Cruz, Vercalos, Versageek, Versus22, Vianello, Vicente Calibo de Jesus, Vicrom, Village Idiot Sabant, VincentHuyTran, Vints, Violetdolor, ViperSnake151, Viperist, Virgil2009, Viriditas, Virlanie, Vivenot, Vladimir ryunin94, VolatileChemical, Volnturista, Vonjobi, Voyagerfan5761, Vssun, WAvegetarian, WJetChao, WWGB, WadeSimMiser, Wahrhaft, Waltloc, Walton One, Watti Renew, Wdls75, Wdmanuel, Websterwebfoot, Welsh, Weregerbil, Wewereright1054, Weyes, Wfaldas, Whereizben, WhisperToMe, Whispering, White Shadows, White aasian, Whkoh, Wierd Jelckama, Wifishark, WikHead, Wiki Raja, Wiki alf, Wikiacc, Wikipaniki, Wikipedia@natividads.com, Wikipedian c, Wikiscribe, Wiknik, Will Beback, Willem Souwer, William Avery, Willy on WheeLs, Wilson44691, Windwarlock, Winhunter, Winjay, Wintersira, Wknight94, Wm asia, Wng, Woodg62, Woohookitty, Workinstruments, Wowzamboangacity, Wtmitchell, Wumingzi, Wxstorm, Wysprgr2005, Xaosflux, Xdenizen, Xelaimpc, Xeltran, Xenovatis, Xiner, Xintata, Xp54321, Xshii, Xtifr, Xtzou, Xyzzy n, YEAHHH JAMIE XP, YZJay, Yamamoto Ichiro, YellowMonkey, YoBobbeh, Youlittlehandsomeguy, Yournecs, Youssefsan, Yuymako, Zahid Abdassabur, Zanter, Zararo, Zeimusu, Zephyr2k, Zerwell, Zhou Yu, Zibem, Zidonuke, Zollerriia, Zscout370, Zsero, ZwickauDeluxe, Île flottante, A, 5597 anonymous edits

Palau *Source*: http://en.wikipedia.org/w/index.php?title=Palau *Contributors*: *drew, 23prootie, 2help, Abhijitsathe, Abstrakt, Academic Challenger, Accuruss, Acntx, AdjustShift, Aec is away, Ageoflo, Ahoerstemeier, Airplaneman, Akanemoto, Albyva, AlefZet, Alinor, Amymiller, Ananas96, Andrwsc, Andy Marchbanks, Andy.goryachev, Angela, Angelo De La Paz, Anthon.Eff, Aotearoa, Aquaimages, Aquaman007, ArgGeo, Arjuna909, ArthurWeasley, Astuishin, Aymatth2, AzNeo, Badagnani, Bart133, Beland, Berry2K, Beyond My Ken, BigBen21, Billinghurst, Biruitorul, Bje2089, Bleubeatle, Bnynms, Bob the Hamster, Bobblehead, Bobblewik, Bobo192, Bouvacio, Bravo Foxtrot, Briaboru, Broncosrballin16, Bryabail, Bryan Derksen, Buaidh, Bullet train, Bumm13, C0nanPayne, C777, Caburianfine, Caerwine, CalicoCatLover, Calmer Waters, Candlewicke, Cantus, CapitalR, Caponer, Chan Han Xiang, Chanheigeorge, Charles Matthews, Chessy999, Chipmunkdavis, ChrisRuvolo, Christian Chenier, Christopher Sundita, Chun-hian, Claverly, Clruscitti, Coemgenus, CommonsDelinker, Conte di Cavour, Conversion script, Corsarius, Credema, Curb Chain, Cybercobra, D6, Dalycitymike, DancingPenguin, Daniscp, Dark Shikari, DarkFireYoshi, Daveswagon, David Kernow, DavideAndrea, Dbabbitt, Dbiel, DeMatt, Dedwarmo, Deipnosophista, Den fjättrade ankan, Denelson83, Dhr.Koster, Dlohcierekim, Dmaestoso, Doc Quintana, DocWatson42, Domaleixo, Douglemeister, Dpaajones, Drbug, Dschroder, Dsinclair.dsaa, Dtobias, Dufekin, Dungodung, Dureo, Dweller, E Pluribus Anthony, Eagle4000, Earl Andrew, Edofmund, Edward, Edward Gow-Smith, Eeekster, EeepEeep, Egon, El C, Electionworld, Elon3523, EncMstr, EncycloPetey, Endroit, Englishfooty, Enzino, Eoghanacht, Epbr123, Errol1968, Euchiasmus, Everyking, Evrik, Fences and windows, Fertejol, Feydey, Fibonacci, Fiftytwo thirty, Filiep, Firebreather3463, Firefliie, Firsfron, Fkhateeb, Flatterworld, Fllblueguy, Fnielsen, Franck Martin, Funnyhat, Gadfium, Gaga930, Gaius Cornelius, Galoubet, Gatta, Geo Swan, Gertjan R., Giraffedata, Glenn, Gogo Dodo, Good Olfactory, Gr8opinionater, Graham87, Grayfm, Green Giant, Grendelkhan, Grutness, Gryffindor, Gt86, Gugganij, GuidoSarducci, HQCentral, Haikon, HappyCamper, Hashar, Hdt83, HeikoEvermann, Henning Blatt, Heron, Hibernian, Hilltop, Himatsu Bushi, HkCaGu, Hmains, Hooperbloob, Horologium, Housemouce, Hppl, Iansal94, IceDragon64, Int21h, Iqinn, Iridescent, Isaac Rabinovitch, Islanesia, JForget, Jakew, JamesAM, Japantvmania, JasonTrue, Jd027, Jeff G., Jeronimo, Jimmy Slade, Jjok, Johammond, Jops87, Joseph Solis in Australia, JoshXF, Jprowland, Jpta, Judeindigo, Jusme1, Just H, KCinDC, Kain Nihil, Kanags, KarlM, Kenatipo, Kennethteh, Ketib37, Kibitzer, Kintetsubuffalo, Kirbygross 123, KirkEN, Kmduffin, Koavf, Kotniski, Koyaanis Qatsi, Kross, Kukini, Kurykh, Kusunose, Kwamikagami, LA2, Laurens-af, LeboKunda, Leo girl90, Lightmouse, Likemac2136, LittleDan, LizardJr8, Ln (-x) = ln x + (pi)i, Locos epraix, Look2See1, Loren.wilton, Luke Maui, Luph25, Lzhhilly, MChew, MER-C, MJCdetroit, MacTire, Magioladitis, Magnet For Knowledge, Malerin, Malhonen, Manchild, Mandarax, Mani1, MarcosFenn, Mariposa111, Markborg, Markcox, Markjohndaley, Markwiki, Martarius, Martinwilke1980, Masem, Matthewprc, McDogm, McSly, Mclay1, Merbabu, Mervyn, Metalman963, Metricator, Metta Bubble, Meyergreen, Mfa fariz, MiLo28, Mic, Michael Dorosh, Mickey gfss2007, Mike Beidler, Miquonranger03, Mitched, Moondyne, Morwen, Moshe Constantine Hassan Al-Silverburg, Movedgood, Mr Tan, MrOllie, Munci, Murgh, Mydotnet, N-true, NLucey, Nach0king, Nakon, Namiba, NatureA16, NawlinWiki, Nbashaw, Nereocystis, Neumannk, Newportm, Ngemlis, Ngewakl, Nice0100, Nick-D, Night w, Nihonjoe, Niteowlneils, Nmpenguin, NorVegan, North King, Notheruser, Nscheffey, Nurg, Ohiostandard, Omdo, Omicronpersei8, Omniplex, Orange Suede Sofa, Orphan Wiki, Osomec, PDH, Packetmonger, Paine Ellsworth, PaintedCarpet, Paki.tv, Pascal, Patrick-br, Paul-L, Pedro Felipe, Persian Poet Gal, Peter E. James, Peter Grey, PeterEasthope, Phantomsteve, Philip Trueman, Physchim62, Piccadilly, Picnic dude, Pigman, Pogo-Pogo-Pogo, Polaron, Polylerus, Popsracer, Postdlf, Primergrey, Pristino, Pruneau, Pschemp, PxT, Pyfan, Qrfqr, R9tgokunks, RBBrittain, Ramirez72, Randyqs, Ratzer, Rennell435, Rich Farmbrough, Rick Block, Rif Winfield, RingtailedFox, Rj1020, Rjoyce, Rjwilmsi, Rkitko, RobNS, Robertpgreer, Roded86400, Rodii, Romanm, Rougarai, Rreagan007, Rune.welsh, Ryanaxp, SMC, Samantha555, Sandstein, Sardanaphalus, Savantpol, Scanlan, Schneelocke, Scipius, Scohoust, ScottDavis, Scottbenbow, Scythian1, Secfan, Sgt Pinback, Shadowjams, Shii, Shlimozzle, SimonP, Sjakkalle, Skookum1, Slawojarek, Sleigh, Snowdog, Snowolf, Snoyes, Someone35, Sophie means wisdom, Souang, Sowsearsoup, Stefan Jansen, StephenDawson, Strangestranger1, Super braulio, Suruena, TFighterPilot, TShilo12, TUF-KAT, Taichi, Takowl, Tamariki, Targeman, Tariqabjotu, Tbhotch, Teacurran, Template namespace initialisation script, The Epopt, The Singing Badger, The Thing That Should Not Be, The Transhumanist, The willwill, The wub, TheFoxdonut, Thecheesykid, Thegreatdr, Thepinkninja2, Therequiembellishere, Thingg, Tim Starling, TorontonianOnlines, TreasuryTag, Twalls, Ucanlookitup, Umybut, Urhixidur, Urotsu, Valentinian, Vardion, Vishnava, Vlma111, VoodooIsland, Vsmith, Waltpohl, Wareq, Washiucho, Wavelength, Wechselstrom, West Brom 4ever, WhisperToMe, Wiki Wikardo, WikiDan61, Wikipeditor, Will Beback, Wkharrisjr, Wmahan, Wmazeditor, Wonkotsane, Woohookitty, Work permit, Wwishnu, XXPatriotXx, Yaromunna, Yeng-Wang-Yeh, Yupik, YuriSapiens, Zanimum, Zdrv, Zippy, Zombie Hunter Smurf, Zorkikat, Zundark, Ævar Arnfjörð Bjarmason, Жељко Тодоровић, , , 581 anonymous edits

Geography_of_Palau *Source*: http://en.wikipedia.org/w/index.php?title=Geography_of_Palau *Contributors*: Amalthea, Angela, Conversion script, D6, Darwinek, Dr. Blofeld, Gaius Cornelius, Hipi Zhdripi, Hmains, Kate, Koyaanis Qatsi, Maximaximax, Pascal, PoccilScript, Sdrawkcab, TreasuryTag, Waltpohl, Warofdreams, 3 anonymous edits

Micronesia *Source*: http://en.wikipedia.org/w/index.php?title=Micronesia *Contributors*: *drew, 000peter, 1297, 23prootie, 43212344321fdsaasdf, A-giau, AKeen, Abhijitsathe, Adrigon, AgadaUrbanit, Ahoerstemeier, Alai, AlexiusHoratius, Alientraveller, Alison9, AlistairMcMillan, Altenmann, Am86, Ambrish.shrivastava, Andre Engels, Andrwsc, Angela, ArchonMeld, Aridd, Arjuna909, Arkuat, Ask123, BK2011, Bardsandwarriors, Beland, Big Adamsky, Biruitorul, Blaylockjam10, Bobo192, Bootstoots, Boy in the bands, Brian Sayrs, Brianski, Bruce Hall, Bsazat, Buaidh, C.m.jones, CJLL Wright, Card, Causa sui, Champo330, Chrisbolt, Civvi, Clarince63, CommonsDelinker, Conversion script, Corvus13, Cycotic, Daarznieks, Danhussey, Danka.Ledgerwood, Daysroach, Demmy, Deror avi, Diligent Terrier, Discospinster, Djsasso, Donarreiskoffer, Dreadstar, Dysprosia, Ebizur, EeepEeep, EncycloPetey, Endroit, Enigmaman, Epbr123, Epistemos, Eugene Banks, Everyking, Evlekis, Fdsaasdffdsaasdf, Fdsaasdffdsaasdffdsaasdf1234, Fireice, Friends77, Gadfium, Gilliam, Gogo Dodo, Gondry, Graham87, Gringo300, Grutness, Gzornenplatz, Hadal, Hairy Dude, Haza-w, Henrygb, Herut, Hibernian, HkCaGu, Hobe, Husond, Ideyal, Influenceuh, J.delanoy, JF Manning, Jekangisher, Joelalvarez617, John K, JohnI, Johnbrownsbody, Jojit fb, Jonkerz, Jonpro, Joseph Dwayne, Juanpdp, Kahuroa, Kanags, Keanu2277, Kentynet, Kgmarsh, Kintetsubuffalo, Knutux, Koavf, Kodster, Koryakov Yuri, Kwamikagami, L Kensington, Lights, LittleOldMe, Llull, Luk, M.O.X, MK8, MacedonianBoy, MadgerBadger, Mangojuice, Mani1, Marek69, Martinwilke1980, Materialscientist, MatthewVanitas, Maximillion Pegasus, Mclay1, Melmann, Merlinme, Michael Hardy, Minimac, Monedula, Moshe Constantine Hassan Al-Silverburg, Moxy, NLucey, NawlinWiki, Nbkkickyou, Nejc.novak, Newportm, Nickjg54, Nigel the, Night w, Nlu, Nmviw, Notheruser, NovaDog, Ntennis, Omicronpersei8, Ondundozonananandana, Patar knight, Paxsimius, PhantomPlugger, Phil Boswell, PlatypeanArchcow, Pras, Prashanthns, Quadell, Quangbao, QuasyBoy, Quizimodo, Qxz, Rarelibra, Raven4x4x, Readin, RedWolf, RepliCarter, Rhythm, Ricky@36, Roastytoast, Rodii, RothePo, Rui Silva, RussBlau, Saintsue28, Sanya3, Scherzomio, Seresin, Shawn in Montreal, Shotwell, Sky Attacker, Soetermans, Spitfirespitfire, Sprinkler21, Stellea, Stephen.peterson, Sterio, Superm401, Superryu, TShilo12, Tcncv, The Transhumanist, The Wild West guy, Thecurran, Thefullback, Thingg, Tide rolls, TonyTheTiger, Tonytam, Trec'hlid mitonet, Trevor MacInnis, Ukexpat, Uncle Dick, Unitedstates1000, Viriditas, Vrenator, Wavelength, WikHead, Wimt, Wing Nut, YeshuaDavid, Yk Yk Yk, Zocky, Zundark, 309 anonymous edits

Island_country *Source*: http://en.wikipedia.org/w/index.php?title=Island_country *Contributors*: Aircorn, Alansohn, Apoivre, Australian Jezza, Avala, Bajanchad, Bendav, Bevo74, Bfigura's puppy, Big Adamsky, Bomac, Branka France, Cacophony, Calabe1992, Ceyockey, Chase me ladies, I'm the Cavalry, Chinaman271, Chipmunkdavis, Chowbok, ChrisHodgesUK, Chubbles, ConCompS, Cursed Pretzel, CurtisSwain, Cyrius, Daarznieks, Danny-w, Dantadd, Davshul, Dondegroovily, Drestros power, Dshfhgsfhadhdsf, Duffman, Dúnadan, Eagle4000, Ebglider91, Ellafb, Epbr123, Episteme-jp, Eumolpo, F-451, Fg2, Frankenpuppy, GEEkEr71, Gilliam, Giraffedata, Green caterpillar, Gzornenplatz, HenryLi, Hm2k, Huaiwei, IJA, IkonicDeath, IngSoc BigBrother, Insanity Incarnate, Iridescent, J 1982, Jahiegel, JamesTseng, Jamezcd, Jao, Jaydec, Jiang, Jjok, Jmundo, Johan Lont, Juliancolton, Julie of the Wolves, Karthikndr, Katieh5584, Katovatzschyn, Kevin Murray, La Pianista, Levineps, LinguistAtLarge, Lowellian, Lulo.it, Lycaon83, Macquarie, ManUnited4Ever, Maurreen, Mendaliv, MiShogun, Midway, Mistakefinder, Mww113, NTK, NiTenIchiRyu, Nick Moss, Northamerica1000, Nubiatech, Nurg, Peridon, Pro66, ProveIt, Python eggs, Qaka, Quintote, Radagast83, Rannpháirtí anaithnid (old), Rapsar, Reuv, RexNL, Rholton, Rjwilmsi, RobertHuaXia, Roux-HG, Rugbyboyrob, SGBailey, SchmuckyTheCat, Segregold, Sentausa, Shanes, SilkTork, SimonP, Smith Jones, Spamwithmustard, Spinach Monster, SpookyMulder, Techdawg667, Teledildonix314, The wub, Tide rolls, Toxicbreakfast, Txof00, Ukabia, Uncle G, Van der Hoorn, VegaDark, Voyevoda, Whytecypress, Wik, Wikipéire, Woodysee, Woohookitty, Wotapalaver, X3210, Xe7al, ZayZayEM, Александър, 157 anonymous edits

Mount_Ngerchelchuus *Source*: http://en.wikipedia.org/w/index.php?title=Mount_Ngerchelchuus *Contributors*: Droll, Grutness, Luke Maui, Racerx11, Rémih, Scheridon, Sl, Thorongilion, WPSamson, Waltloc, 2 anonymous edits

Babeldaob *Source*: http://en.wikipedia.org/w/index.php?title=Babeldaob *Contributors*: Andrevruas, Bardsandwarriors, Btw, Colonies Chris, CristianoMacaluso, Danny, Epp, Eriknjorl, Gertjan R., Graeme374, Grutness, Gugganij, Hmains, Joel7687, Jon Harald Søby, Jpo, Man-made Maiden, MatthewVanitas, Nurg, Poulpy, Ratzer, Rennell435, Sl, TShilo12, Taichi, Timo Honkasalo, TrinityGate, Waltpohl, WhisperToMe, Wilchett, XLR8TION, Zoicon5, 17 anonymous edits

Environment_of_Palau *Source*: http://en.wikipedia.org/w/index.php?title=Environment_of_Palau *Contributors*: Alan Liefting, Thinking of England, 2 anonymous edits

Wildlife_of_Palau *Source*: http://en.wikipedia.org/w/index.php?title=Wildlife_of_Palau *Contributors*: Night w

States_of_Palau *Source*: http://en.wikipedia.org/w/index.php?title=States_of_Palau *Contributors*: Acntx, Amikake3, Anonymous Dissident, Antipoeten, Aotearoa, CUSENZA Mario, Chanheigeorge, Crusoe8181, D6, Darwinek, Docu, Electionworld, Ginga123, Good Olfactory, Hmains, Kintetsubuffalo, KirkEN, Luke Maui, NERIUM, NL-Ninane, PDH, Poulpy, Rarelibra, Ratzer, Rif Winfield, Tobias Conradi, Tomchiukc, Trilobite, VirtualDelight, 12 anonymous edits

Demographics_of_Palau *Source*: http://en.wikipedia.org/w/index.php?title=Demographics_of_Palau *Contributors*: Blamed, CUSENZA Mario, Conversion script, D6, FargomeD, Gugganij, Hmains, Koyaanis Qatsi, Kwamikagami, Losau12345, Mangostar, Neelix, Nihonjoe, Ratzer, SimonP, Targovishtenec bg, The Transhumanist, Thehelpfulone, TreasuryTag, Waltpohl, 6 anonymous edits

Melekeok *Source*: http://en.wikipedia.org/w/index.php?title=Melekeok *Contributors*: Adimovk5, Anthony Appleyard, Aotearoa, BrunoSR, Chanheigeorge, Gertjan R., Johnpatona, Jéské Couriano, Luke Maui, Rémih, Sarcelles, Severo, Thylacinus cynocephalus, WhisperToMe, Мл.научный сотрудник, 2 anonymous edits

Politics_of_Palau *Source*: http://en.wikipedia.org/w/index.php?title=Politics_of_Palau *Contributors*: -- April, Academic Challenger, Acntx, AldenWeer, CDN99, CalJW, Conversion script, D6, Danny, Electionworld, Hadal, HopeSeekr of xMule, Katefan0, Kintetsubuffalo, Koyaanis Qatsi, Mbakkel2, Nightstallion, Niteowlneils, Patangay, Rich Farmbrough, Richard75, RickK, Shiro-gi, Snoyes, Soundvisions1, Steve espinola, The Transhumanist, Tlaresch, TracyRenee, User2004, Warofdreams, Wilfried Derksen, 9 anonymous edits

Oceania *Source*: http://en.wikipedia.org/w/index.php?title=Oceania *Contributors*: *drew, 08OceanBeach SD, 1549bcp, 203.109.250.xxx, 23prootie, 2toy mora, ;';'iH Hi';';, A dumb computer, A the 0th, A-giau, ANZUS, ASDFGH, Abhijitsathe, Ace ETP, Acs4b, AdamSommerton, Aeonx, Aesopos, Ahills60, Ahoerstemeier, Airplaneman, Aitias, AkBowl, Akanemoto, Akuindo, Alai, Alansohn, AlexCovarrubias, Alice, Alinor, Alphathon, Altermike, Alton.arts, Ancheta Wis, Andre Engels, Andrejj, Andrelvis, Andres, Andrwsc, Andyjsmith, Angelo De La Paz, Animum, AnnaFrance, AnnaJGrant, Arakunem, Aris Katsaris, ArnoldPlaton, Asmalik94, Australia boy, Avenue, AxelBoldt, BRG, Bazonka, Bcnviajero, Bduke, Before My Ken, Beland, Ben Arnold, Ben Kidwell, Bencollins, Bkell, Blaylockjam10, Bluedevil14, Bobo192, Bongwarrior, Bosonic dressing, Bowloca, Bradv, Brian, Brian0918, Brianga, Buaidh, Bugoslav, CJ, CJLL Wright, Caltas, Can't sleep, clown will eat me, Canadian-Bacon, Cantthinkofoneespn, Cantus, CeeWhy2, Cenarium, Chanheigeorge, Chipmunkdavis, Chode78, Ckatz, Cmdrjameson, Coemgenus, Cogito ergo sumo, Connormah, Conversion script, Corticopia, Crazygraham, Creol, Cricketman274, Crissov, Crusadeonilliteracy, Curps, Cursive, Cyan, DJ Clayworth, DMacks, Daarznieks, Dabomb87, Daniel Olsen, DanielCD, Dave420, David Kernow, David0811, Dayv, Ddude700, Dentren, DerHexer, Dfrg.msc, Dhum Dhum, Dinesh smita, Discospinster, Dmmaus, Dn9ahx, Donnermonkey, Doseiai2, Doug Bell, Dpark, Dr. Blofeld, Dwo, E Pluribus Anthony, EJF, Ebizur, El C, Elbac14, Elonka, EmmetCaulfield, Enzino, Eric-Wester, Espetkov, Eu.stefan, Everyking, Excirial, Fabricationary, Factoid Killer, Felix Portier, FlaggyAznavour, FlavrSavr, Florm, Flyingidiot, Foobarnix, Frecklefoot, Frehley, Funandtrvl, Fuzheado, G_from_B, Gadfium, Gaimhreadhan, Galactiger, Gelpiac, Gergyl, Gfcvoice, Gilgamesh, Girmitya, Glenn, Gnangarra, Go for it!, Goldfishbutt, Green caterpillar, Gridlocked, Grillo, Grumbular Shmo, Grunners, Grutness, Gurubrahma, Guy Peters, Hack, Harryboyles, Heff01, HeikoEvermann, Henry W. Schmitt, Hmains, Hobe, Hugo999, Humphrey Jungle, Husond, Influenceuh, Issie12345, Ivirivi00, Iwanttoeditthissh, J'onn J'onzz, J.delanoy, JMK, JackofOz, Jacob.jose, James Balti, JamesR, JamesyWamesy, Jasper33, Javierito92, Jcmenal, Jeffq, Jerryseinfeld, Jfitts, Jfpierce, Jimothytrotter, Jimp, John Riemann Soong, JohnCman, Jonkerz, Jose77, Jpta, Jraytram, Jrockley, Juanediaz, Jumbuck, Jusjih, Just James, K, Kahuroa, Kalogeropoulos, Kentynet, Khajidha, Kimjae20, King kong922, Kingpin13, Kirbygross 123, Klippa, Knutux, Koavf, Kontos, Krellis, Kudarat, Kuru, Kwamikagami, L33th4x0rguy, Largoplazo, Latitude0116, Leafyplant, Lightmouse, Limegreen, Linmhall, Liopa, Little muddy funkster, Loggong, Look2See1, Loopy, Lotje, Luk, Luna Santin, M.O.X, MC10, MCSparki, MPerel, MaCRoEco, Mac, Macarenses, Malhonen, Maloti090, Man vyi, Mani1, Mannafredo, Mannerheim, Mareino, Marek69, Marianocecowski, Mariusmiti, Mark, Martarius, MaryPoppins878, Matt8998, Maulucioni, Maurreen, Mav, Mayumashu, McDogm, Mclay1, Megan1967, Mendalus, Merbabu, Merovingian, Mhicaoidh, MiKHEILL, Mikaey, Mike Beidler, Minervamoon, Mmggggmgodl, Mnmazur, Monedula, Monkeyman, Montrealais, Moriori, Moverton, Mozasaur, Mr Bungle, Mrdice, Muijz, Munci, Musicpvm, Mwanner, Mwdalton, Mxn, My76Strat, Mygerardromance, Mynci2894, Nagy, Nationalgeo, NawlinWiki, NeoJustin, Neverquick, Newportm, Nick UA, NickFeder, Night w, Ninly, NoPuzzleStranger, Northamerica1000, Notscott, Nurg, Nyttend, OffsBlink, Onemore9, Oreo Priest, Otherone, Owen82100, PRSky, Pacian, Paine Ellsworth, Patrick, Paul Stansifer, Paul foord, Peaceray, PhJ, Philip Trueman, Philopp, Pietro, Pne, Polar bear, Porqin, Postdlf, Prolog, Python eggs, QmunkE, Quintucket, Quizimodo, R'n'B, R9tgokunks, RCIWesner, RG104, Ratzer, RexNL, Rich Farmbrough, Richard Weil, Risker, Rjwilmsi, Rodii, RoyBoy, Rulesfan, Runefrost, Russavia, S3000, SGGH, SUPERGRUNT, Saberwolf116, Saberwyn, SaintDaveUK, Salt Yeung, Samwingkit, Sanyo Nokia, Sarpasarpeti101, SatuSuro, ScottMainwaring, ScottyMull, Sebasbronzini, Serp38021, Sfahey, Shmilyshy, Shushruth, Sideshow Bob Roberts, SimonLyall, Sir Tanx, Sleigh, Sliat 1981, Slightsmile, Smithalexer, Snori, Some jerk on the Internet, Somebody in the WWW, SpacemanSpiff, Sportskiddwi1, Srnec, Sss333, StaticGull, Stephen, Stephenb, StillWalking, Supaman89, Sven Eriksson, TFOWR, TShilo12, Taifarious1, TakuyaMurata, Tannin, Tbhotch, Teapotgeorge, Teinesavaii, Template namespace initialisation script, Th1rt3en, The Singing Badger, The Thing That Should Not Be, The Transhumanist, TheGreerFamily, TheNightFox1981, Thedoctor98, Thehookemanrevolution, Thephenom92, Therequiembellishere, Thingg, Tide rolls, Tiki Nss, Timir Saxa, Tins128, Tobias Conradi, Tom Radulovich, Tpbradbury, TracySurya, Twas Now, Tzartzam, Ukexpat, Uncreated, Unitedstates1000, Uxejn, VDK, Vanibal2, Vicki Rosenzweig, Victoria Clark 86, Vsmith, Vuong Ngan Ha, WTFPWNER2, Wakablogger2, Wantok, Wareh, WereSpielChequers, Western Pines, WikiParker, Wikipedico wikipedico, Wildthing61476, Will Beback, Windharp, Winstonwolfe, Wiz9999, Worldtimezone, Wprlh, Wtmitchell, Wwahammy, X3210, XavierGreen, Xpesten, Yacht, Yamamoto Ichiro, Youandme, ZanderSchubert, Zigger, Zoe, ZwickauDeluxe, Ævar Arnfjörð Bjarmason, لی قع فش ا, 759 anonymous edits

President_of_Palau *Source*: http://en.wikipedia.org/w/index.php?title=President_of_Palau *Contributors*: *drew, Acntx, Akuindo, Awbeal, Biruitorul, Crusoe8181, Hemlock Martinis, Hmains, Jaba1977, Joao Xavier, Lightmouse, NYYCF, Salamurai, Scanlan, Severo, Snappy, Snocrates, Sundostund, Therequiembellishere, 7 anonymous edits

House_of_Delegates *Source*: http://en.wikipedia.org/w/index.php?title=House_of_Delegates *Contributors*: Boffin, Dale Arnett, Edward, Ground, JII, Malcolma, MisfitToys, Park3r, Sandman30s, Spidern, Sumsum2010, 4 anonymous edits

Foreign_relations_of_Palau *Source*: http://en.wikipedia.org/w/index.php?title=Foreign_relations_of_Palau *Contributors*: Alan Liefting, Auntof6, D6, Electionworld, Henning Blatt, Hmains, Hwliang, Instantnood, Jiang, Koyaanis Qatsi, Livajo, MarcosFenn, Severo, Super Mario, 12 anonymous edits

History_of_Palau *Source*: http://en.wikipedia.org/w/index.php?title=History_of_Palau *Contributors*: 62.253.64.xxx, Andrewa, CUSENZA Mario, CalJW, Chrishmt0423, Conversion script, D6, Darwinek, DocWatson42, Edward, Filiocht, Fuzheado, Gaius Cornelius, Gzornenplatz, Hmains, Howcheng, J87j8787, Ketib37, Kingpin13, KirkEN, Koyaanis Qatsi, Kwamikagami, Legis, Ohconfucius, Omnipaedista, RedWolf, Rich Farmbrough, Roke, Sam Blacketer, Sesel, SimonArlott, SimonP, Sisyph, Toya, TrinityGate, Warofdreams, WikHead, Wikilibrarian, 5 anonymous edits

Economy_of_Palau *Source*: http://en.wikipedia.org/w/index.php?title=Economy_of_Palau *Contributors*: Biruitorul, Bobblewik, CUSENZA Mario, Camembert, D6, Extra999, Flyingtoaster1337, HQCentral, Hmains, Jiang, Kbdank71, Kennethteh, Koyaanis Qatsi, Liveste, Moon&Nature, Rich Farmbrough, Roadwarrior52, Sebleouf, The Epopt, 14 anonymous edits

Image Sources, Licenses and Contributors

Image:LocationPalau.png *Source*: http://en.wikipedia.org/w/index.php?title=File:LocationPalau.png *License*: unknown *Contributors*: User:Vardion

Image:Flag of Palau.svg *Source*: http://en.wikipedia.org/w/index.php?title=File:Flag_of_Palau.svg *License*: unknown *Contributors*: User:Gabbe

Image:Seal_of_Palau.png *Source*: http://en.wikipedia.org/w/index.php?title=File:Seal_of_Palau.png *License*: unknown *Contributors*: Mononomic (talk) Original uploader was Mononomic at en.wikipedia

Image:Un-palau.png *Source*: http://en.wikipedia.org/w/index.php?title=File:Un-palau.png *License*: unknown *Contributors*: Jeroencommons

Image:Palau-CIA WFB Map.png *Source*: http://en.wikipedia.org/w/index.php?title=File:Palau-CIA_WFB_Map.png *License*: unknown *Contributors*: Benchill, Hoshie, Jonathan Harker, Ras67, Spangineer, Wolfman

Image:Peleliu map.jpg *Source*: http://en.wikipedia.org/w/index.php?title=File:Peleliu_map.jpg *License*: unknown *Contributors*: USGS

Image:Helen Reef-NASA.png *Source*: http://en.wikipedia.org/w/index.php?title=File:Helen_Reef-NASA.png *License*: unknown *Contributors*: BaldBoris, Ratzer, Telim tor

Image:States of Palau.jpg *Source*: http://en.wikipedia.org/w/index.php?title=File:States_of_Palau.jpg *License*: unknown *Contributors*: Kirk, 2 anonymous edits

File:Small Flag of the United Nations ZP.svg *Source*: http://en.wikipedia.org/w/index.php?title=File:Small_Flag_of_the_United_Nations_ZP.svg *License*: unknown *Contributors*: Collard, Homo lupus, Joowwww, Petar Marjanovic, Sceptic, XcepticZP, 1 anonymous edits

File:UN Trusteeship Council.jpg *Source*: http://en.wikipedia.org/w/index.php?title=File:UN_Trusteeship_Council.jpg *License*: unknown *Contributors*: J.reed at en.wikipedia

File:Flag of France.svg *Source*: http://en.wikipedia.org/w/index.php?title=File:Flag_of_France.svg *License*: unknown *Contributors*: Anomie

File:Flag of the United Kingdom.svg *Source*: http://en.wikipedia.org/w/index.php?title=File:Flag_of_the_United_Kingdom.svg *License*: unknown *Contributors*: Anomie, Mifter

File:Decolonization - World In 1945 en.svg *Source*: http://en.wikipedia.org/w/index.php?title=File:Decolonization_-_World_In_1945_en.svg *License*: unknown *Contributors*: United Nations Department of Field Support, Cartographic Section

File:Decolonization - World Today en.svg *Source*: http://en.wikipedia.org/w/index.php?title=File:Decolonization_-_World_Today_en.svg *License*: unknown *Contributors*: United Nations Department of Field Support, Cartographic Section

File:TTPI UN Mission 1978.jpg *Source*: http://en.wikipedia.org/w/index.php?title=File:TTPI_UN_Mission_1978.jpg *License*: unknown *Contributors*: CarolSpears, Infrogmation, J o, Joowwww, WhisperToMe, , 1 anonymous edits

File:Flag of the Philippines.svg *Source*: http://en.wikipedia.org/w/index.php?title=File:Flag_of_the_Philippines.svg *License*: unknown *Contributors*: Aira Cutamora

File:Coat of Arms of the Philippines.svg *Source*: http://en.wikipedia.org/w/index.php?title=File:Coat_of_Arms_of_the_Philippines.svg *License*: unknown *Contributors*: User:Zscout370

File:The Philippines and ASEAN (orthographic projection).svg *Source*: http://en.wikipedia.org/w/index.php?title=File:The_Philippines_and_ASEAN_(orthographic_projection).svg *License*: unknown *Contributors*: User:Connormah

File:Increase2.svg *Source*: http://en.wikipedia.org/w/index.php?title=File:Increase2.svg *License*: unknown *Contributors*: Sarang

Image:Speakerlink.svg *Source*: http://en.wikipedia.org/w/index.php?title=File:Speakerlink.svg *License*: unknown *Contributors*: Woodstone. Original uploader was Woodstone at en.wikipedia

File:Boxer codex.jpg *Source*: http://en.wikipedia.org/w/index.php?title=File:Boxer_codex.jpg *License*: unknown *Contributors*: Gyzmr, 1 anonymous edits

File:Filipino Ilustrados Jose Rizal Marcelo del Pilar Mariano Ponce.jpg *Source*: http://en.wikipedia.org/w/index.php?title=File:Filipino_Ilustrados_Jose_Rizal_Marcelo_del_Pilar_Mariano_Ponce.jpg *License*: unknown *Contributors*: Original uploader was Lynyrd Salapare at en.wikipedia

File:edsa shrine.jpg *Source*: http://en.wikipedia.org/w/index.php?title=File:Edsa_shrine.jpg *License*: unknown *Contributors*: athos11 from mandaluyong, philippines

File:BRP Goyo PASSEX July2011 2.jpg *Source*: http://en.wikipedia.org/w/index.php?title=File:BRP_Goyo_PASSEX_July2011_2.jpg *License*: unknown *Contributors*: U.S. Navy / Mass Communication Specialist Seaman Sean Furey

File:Embassy of the Philippines, Washington, D.C..jpg *Source*: http://en.wikipedia.org/w/index.php?title=File:Embassy_of_the_Philippines,_Washington,_D.C..jpg *License*: unknown *Contributors*: Asiir at en.wikipedia (Aaron Siirila)

File:Ph regions and provinces.png *Source*: http://en.wikipedia.org/w/index.php?title=File:Ph_regions_and_provinces.png *License*: unknown *Contributors*: User:seav

File:Pana Banaue Rice Terraces.jpg *Source*: http://en.wikipedia.org/w/index.php?title=File:Pana_Banaue_Rice_Terraces.jpg *License*: unknown *Contributors*: Armigo, KTo288, Ori~, 1 anonymous edits

File:Magnify-clip.png *Source*: http://en.wikipedia.org/w/index.php?title=File:Magnify-clip.png *License*: unknown *Contributors*: User:Erasoft24

File:Bohol Tarsier.jpg *Source*: http://en.wikipedia.org/w/index.php?title=File:Bohol_Tarsier.jpg *License*: unknown *Contributors*: mtoz

File:Super Typhoon Megi Oct 18 2010.jpg *Source*: http://en.wikipedia.org/w/index.php?title=File:Super_Typhoon_Megi_Oct_18_2010.jpg *License*: unknown *Contributors*: NASA

File:Makati skyline mjlsha.jpg *Source*: http://en.wikipedia.org/w/index.php?title=File:Makati_skyline_mjlsha.jpg *License*: unknown *Contributors*: mjlsha

File:Monument To Immortality.jpg *Source*: http://en.wikipedia.org/w/index.php?title=File:Monument_To_Immortality.jpg *License*: unknown *Contributors*: Bluemask, Stalmannen, 3 anonymous edits

File:Philippines-demography.png *Source*: http://en.wikipedia.org/w/index.php?title=File:Philippines-demography.png *License*: unknown *Contributors*: http://faostat.fao.org/

File:Philippine ethnic groups per province.PNG *Source*: http://en.wikipedia.org/w/index.php?title=File:Philippine_ethnic_groups_per_province.PNG *License*: unknown *Contributors*: User:Howard the Duck

File:WTMP Pangkat E-14-1.JPG *Source*: http://en.wikipedia.org/w/index.php?title=File:WTMP_Pangkat_E-14-1.JPG *License*: unknown *Contributors*: Stephanie Azarias & Alexander Olpindo (Wikipedia Takes Manila participant)

File:NAIA Terminal 3 2009 MC.jpg *Source*: http://en.wikipedia.org/w/index.php?title=File:NAIA_Terminal_3_2009_MC.jpg *License*: unknown *Contributors*: Mithril Cloud. Original uploader was Mithril Cloud at en.wikipedia

File:Kulintang 06.jpg *Source*: http://en.wikipedia.org/w/index.php?title=File:Kulintang_06.jpg *License*: unknown *Contributors*: PhilipDM, Skipjack

File:Barasoain church ~MVI~ (gaga over Mondo Marcos).jpg *Source*: http://en.wikipedia.org/w/index.php?title=File:Barasoain_church_~MVI~_(gaga_over_Mondo_Marcos).jpg *License*: unknown *Contributors*: ~MVI~ (gaga over Mondo Marcos)

File:Halo halo1.jpg *Source*: http://en.wikipedia.org/w/index.php?title=File:Halo_halo1.jpg *License*: unknown *Contributors*: Dodo, Viriditas

File:Araneta Coliseum Basketball with Big Cube 2011.JPG *Source*: http://en.wikipedia.org/w/index.php?title=File:Araneta_Coliseum_Basketball_with_Big_Cube_2011.JPG *License*: unknown *Contributors*: User:WayKurat

File:Flag of Palau.svg *Source*: http://en.wikipedia.org/w/index.php?title=File:Flag_of_Palau.svg *License*: unknown *Contributors*: User:Gabbe

File:Seal_of_Palau.svg *Source*: http://en.wikipedia.org/w/index.php?title=File:Seal_of_Palau.svg *License*: unknown *Contributors*: Washiucho

File:Palau.png *Source*: http://en.wikipedia.org/w/index.php?title=File:Palau.png *License*: unknown *Contributors*: VoodooIsland, 1 anonymous edits

File:Speaker Icon.svg *Source*: http://en.wikipedia.org/w/index.php?title=File:Speaker_Icon.svg *License*: unknown *Contributors*: Blast, G.Hagedorn, Mobius, 2 anonymous edits

Image:Palau-rock-islands20071222.jpg *Source*: http://en.wikipedia.org/w/index.php?title=File:Palau-rock-islands20071222.jpg *License*: unknown *Contributors*: Peter R. Binter (=Binter); Original uploader was Binter at de.wikipedia (Original text : Peter Binter)

Image:WCTC.JPG *Source*: http://en.wikipedia.org/w/index.php?title=File:WCTC.JPG *License*: unknown *Contributors*:

File:Aerial view limestone islands palau1971.jpg *Source*: http://en.wikipedia.org/w/index.php?title=File:Aerial_view_limestone_islands_palau1971.jpg *License*: unknown *Contributors*: Dr. James P. McVey, NOAA Sea Grant Program

Image:Japan Palau Friendship Bridge.jpg *Source*: http://en.wikipedia.org/w/index.php?title=File:Japan_Palau_Friendship_Bridge.jpg *License*: unknown *Contributors*: Original uploader was Randyqs at en.wikipedia

Image:Desertisland.jpg *Source*: http://en.wikipedia.org/w/index.php?title=File:Desertisland.jpg *License*: unknown *Contributors*: Dr. James P. McVey, NOAA Sea Grant Program

Image:Survivor.palau.logo.png *Source*: http://en.wikipedia.org/w/index.php?title=File:Survivor.palau.logo.png *License*: unknown *Contributors*: AdamJacobMuller, Reywas92, Tom Morris, Untitledmind72

Image:Palau and oceania.jpg *Source*: http://en.wikipedia.org/w/index.php?title=File:Palau_and_oceania.jpg *License*: unknown *Contributors*: Deadstar, Kintetsubuffalo, Ras67, TUBS

File:Palau map.png *Source*: http://en.wikipedia.org/w/index.php?title=File:Palau_map.png *License*: unknown *Contributors*: Modified by Dr. Blofeld

Image:Pacific Culture Areas.jpg *Source*: http://en.wikipedia.org/w/index.php?title=File:Pacific_Culture_Areas.jpg *License*: unknown *Contributors*: User:Kahuroa

Image:Micronesian Cultural Area.png *Source*: http://en.wikipedia.org/w/index.php?title=File:Micronesian_Cultural_Area.png *License*: unknown *Contributors*: User:Kahuroa

File:Flag of Kiribati.svg *Source*: http://en.wikipedia.org/w/index.php?title=File:Flag_of_Kiribati.svg *License*: unknown *Contributors*: User:SKopp

File:Flag of Guam.svg *Source*: http://en.wikipedia.org/w/index.php?title=File:Flag_of_Guam.svg *License*: unknown *Contributors*: User:Denelson83

File:Flag of the Northern Mariana Islands.svg *Source*: http://en.wikipedia.org/w/index.php?title=File:Flag_of_the_Northern_Mariana_Islands.svg *License*: unknown *Contributors*: BanyanTree, Editor at Large, Homo lupus, Liftarn, Mattes, Nightstallion, 3 anonymous edits

File:Flag of the Marshall Islands.svg *Source*: http://en.wikipedia.org/w/index.php?title=File:Flag_of_the_Marshall_Islands.svg *License*: unknown *Contributors*: User:SKopp

File:Flag of Federated States of Micronesia.svg *Source*: http://en.wikipedia.org/w/index.php?title=File:Flag_of_Federated_States_of_Micronesia.svg *License*: unknown *Contributors*: -

File:Flag of Nauru.svg *Source*: http://en.wikipedia.org/w/index.php?title=File:Flag_of_Nauru.svg *License*: unknown *Contributors*: User:SKopp

File:Flag of the United States.svg *Source*: http://en.wikipedia.org/w/index.php?title=File:Flag_of_the_United_States.svg *License*: unknown *Contributors*: Anomie

Image:Island nations.svg *Source*: http://en.wikipedia.org/w/index.php?title=File:Island_nations.svg *License*: unknown *Contributors*: User:Slomox

Image:Borderless countries.PNG *Source*: http://en.wikipedia.org/w/index.php?title=File:Borderless_countries.PNG *License*: unknown *Contributors*: User:Roke

Image:Mount Ngerchelchuus.JPG *Source*: http://en.wikipedia.org/w/index.php?title=File:Mount_Ngerchelchuus.JPG *License*: unknown *Contributors*: U s e r : , U s e r :

file:Palau location map.svg *Source*: http://en.wikipedia.org/w/index.php?title=File:Palau_location_map.svg *License*: unknown *Contributors*: NordNordWest

Image:Beach-in-melekeok20071222.jpg *Source*: http://en.wikipedia.org/w/index.php?title=File:Beach-in-melekeok20071222.jpg *License*: unknown *Contributors*: Peter R. Binter (=Binter); Original uploader was Binter at de.wikipedia (Original text : Peter R. Binter)

Image:State of Melekeok.png *Source*: http://en.wikipedia.org/w/index.php?title=File:State_of_Melekeok.png *License*: unknown *Contributors*: User:Aotearoa

Image:Flag of Melekeok.png *Source*: http://en.wikipedia.org/w/index.php?title=File:Flag_of_Melekeok.png *License*: unknown *Contributors*: Jean Körp, Juergenk59, Ninane

Image:Melekeok.png *Source*: http://en.wikipedia.org/w/index.php?title=File:Melekeok.png *License*: unknown *Contributors*: Aotearoa, Gugganij, Jean Körp

Image:Palau Capitol Dec 2006.jpg *Source*: http://en.wikipedia.org/w/index.php?title=File:Palau_Capitol_Dec_2006.jpg *License*: unknown *Contributors*: Shiro-gi

Image:Executive Seal of Palau.png *Source*: http://en.wikipedia.org/w/index.php?title=File:Executive_Seal_of_Palau.png *License*: unknown *Contributors*: Original uploader was Kintetsubuffalo at en.wikipedia Original uploader was Mononomic at en.wikipedia

File:Orthographic projection centred over Wallis and Futuna Islands.png *Source*: http://en.wikipedia.org/w/index.php?title=File:Orthographic_projection_centred_over_Wallis_and_Futuna_Islands.png *License*: unknown *Contributors*: Electionworld, Geo Swan, KJG2007, Olivier2, Telim tor

Image:Oceania.jpg *Source*: http://en.wikipedia.org/w/index.php?title=File:Oceania.jpg *License*: unknown *Contributors*: Beland

File:Oceania (orthographic projection).svg *Source*: http://en.wikipedia.org/w/index.php?title=File:Oceania_(orthographic_projection).svg *License*: unknown *Contributors*: User:Ch1902

File:Oceania Political Map (EEZ based).png *Source*: http://en.wikipedia.org/w/index.php?title=File:Oceania_Political_Map_(EEZ_based).png *License*: unknown *Contributors*: w:en:User:AlinorAlinor (w:en:User talk:Alinortalk)

Image:Oceania_satellite.jpg *Source*: http://en.wikipedia.org/w/index.php?title=File:Oceania_satellite.jpg *License*: unknown *Contributors*: AGoon,

File:Oceania, broad (orthographic projection).svg *Source*: http://en.wikipedia.org/w/index.php?title=File:Oceania,_broad_(orthographic_projection).svg *License*: unknown *Contributors*: User:rohith_goura

File:Oceania, narrow (orthographic projection).svg *Source*: http://en.wikipedia.org/w/index.php?title=File:Oceania,_narrow_(orthographic_projection).svg *License*: unknown *Contributors*: User:Kwamikagami

File:Map OC-Oceania.PNG *Source*: http://en.wikipedia.org/w/index.php?title=File:Map_OC-Oceania.PNG *License*: unknown *Contributors*: User:Hobe

File:Flag of Australia.svg *Source*: http://en.wikipedia.org/w/index.php?title=File:Flag_of_Australia.svg *License*: unknown *Contributors*: Anomie, Mifter

File:Flag of New Zealand.svg *Source*: http://en.wikipedia.org/w/index.php?title=File:Flag_of_New_Zealand.svg *License*: unknown *Contributors*: Adambro, Arria Belli, Avenue, Bawolff, Bjankuloski06en, ButterStick, Denelson83, Donk, Duduziq, EugeneZelenko, Fred J, Fry1989, Hugh Jass, Ibagli, Jusjih, Klemen Kocjancic, Mamndassan, Mattes, Nightstallion, O, Peeperman, Poromiami, Reisio, Rfc1394, Shizhao, Tabasco, Transparent Blue, Väsk, Xufanc, Zscout370, 35 anonymous edits

File:Flag of Christmas Island.svg *Source*: http://en.wikipedia.org/w/index.php?title=File:Flag_of_Christmas_Island.svg *License*: unknown *Contributors*: Anime Addict AA, Denelson83, Docu, Erribas, Fry1989, Homo lupus, Hoshie, Mattes, Ninane, TFCforever, Telim tor, Wknight94, Xufanc, Zscout370, 1 anonymous edits

File:Flag of the Cocos (Keeling) Islands.svg *Source*: http://en.wikipedia.org/w/index.php?title=File:Flag_of_the_Cocos_(Keeling)_Islands.svg *License*: unknown *Contributors*: AFBorchert, Anime Addict AA, AnonMoos, Denelson83, Fry1989, Homo lupus, Mattes, Ninane, TFCforever, Telim tor, Urhixidur, Xufanc, 1 anonymous edits

File:Flag of Norfolk Island.svg *Source*: http://en.wikipedia.org/w/index.php?title=File:Flag_of_Norfolk_Island.svg *License*: unknown *Contributors*: Avala, Bapho, ChongDae, David Kernow, Dbenbenn, Denniss, George McFinnigan, Himasaram, Homo lupus, Nightstallion, Ninane, Pumbaa80, Slomox, Telim tor, Zscout370, 2 anonymous edits

File:Flag of Fiji.svg *Source*: http://en.wikipedia.org/w/index.php?title=File:Flag_of_Fiji.svg *License*: unknown *Contributors*: Anime Addict AA, Avala, ButterStick, Denelson83, Fred the Oyster, Fry1989, Greentubing, Herbythyme, Homo lupus, Klemen Kocjancic, Krun, Lokal Profil, Mattes, Multichill, Neq00, Nightstallion, Sam916, Urhixidur, Vzb83, 8 anonymous edits

File:Flag of New Caledonia.svg *Source*: http://en.wikipedia.org/w/index.php?title=File:Flag_of_New_Caledonia.svg *License*: unknown *Contributors*: User:Urmas

File:Flag of Papua New Guinea.svg *Source*: http://en.wikipedia.org/w/index.php?title=File:Flag_of_Papua_New_Guinea.svg *License*: unknown *Contributors*: User:Nightstallion

File:Flag of the Solomon Islands.svg *Source*: http://en.wikipedia.org/w/index.php?title=File:Flag_of_the_Solomon_Islands.svg *License*: unknown *Contributors*: User:SKopp

File:Flag of Vanuatu.svg *Source*: http://en.wikipedia.org/w/index.php?title=File:Flag_of_Vanuatu.svg *License*: unknown *Contributors*: Alkari, EugeneZelenko, Fry1989, Homo lupus, Klemen Kocjancic, Mattes, Mikiofpersia, Neq00, Nightstallion, OAlexander, PeterSymonds, Vzb83

File:Flag of Wake Island.svg *Source*: http://en.wikipedia.org/w/index.php?title=File:Flag_of_Wake_Island.svg *License*: unknown *Contributors*: User:Denelson83

File:Flag of American Samoa.svg *Source*: http://en.wikipedia.org/w/index.php?title=File:Flag_of_American_Samoa.svg *License*: unknown *Contributors*: by Tobias Jakobs

File:Flag of the Cook Islands.svg *Source*: http://en.wikipedia.org/w/index.php?title=File:Flag_of_the_Cook_Islands.svg *License*: unknown *Contributors*: User:Denelson83, User:DiamondVertex

File:Flag of Rapa Nui, Chile.svg *Source*: http://en.wikipedia.org/w/index.php?title=File:Flag_of_Rapa_Nui,_Chile.svg *License*: unknown *Contributors*: Mario Fuente Cid, Done on Inkscape Free Software

File:Flag of French Polynesia.svg *Source*: http://en.wikipedia.org/w/index.php?title=File:Flag_of_French_Polynesia.svg *License*: unknown *Contributors*: User:Nightstallion

File:Flag of Hawaii.svg *Source*: http://en.wikipedia.org/w/index.php?title=File:Flag_of_Hawaii.svg *License*: unknown *Contributors*: Anime Addict AA, Awg1010, Dbenbenn, Denelson83, Dzordzm, Editor at Large, F. F. Fjodor, Fry1989, Homo lupus, Hydrargyrum, Idh0854, KAVEBEAR, Kalathalan, Manuelt15, Mattes, Nagy, Nightstallion, Serinde, Svgalbertian, Telim tor, Vonvon, Yaddah, Zscout370, 17 anonymous edits

File:Flag of Niue.svg *Source*: http://en.wikipedia.org/w/index.php?title=File:Flag_of_Niue.svg *License*: unknown *Contributors*: User:SKopp

File:Flag of the Pitcairn Islands.svg *Source*: http://en.wikipedia.org/w/index.php?title=File:Flag_of_the_Pitcairn_Islands.svg *License*: unknown *Contributors*: User:Dbenbenn

File:Flag of Samoa.svg *Source*: http://en.wikipedia.org/w/index.php?title=File:Flag_of_Samoa.svg *License*: unknown *Contributors*: Alkari, Anime Addict AA, Avala, Emijrp, Fry1989, Homo lupus, Klemen Kocjancic, Kookaburra, Kronocide, Mattes, Nightstallion, TFCforever, Telim tor, Vzb83, Xufanc, Zscout370, 1 anonymous edits

File:Flag of Tokelau.svg *Source*: http://en.wikipedia.org/w/index.php?title=File:Flag_of_Tokelau.svg *License*: unknown *Contributors*: User:Andrwsc

File:Flag of Tonga.svg *Source*: http://en.wikipedia.org/w/index.php?title=File:Flag_of_Tonga.svg *License*: unknown *Contributors*: Badseed, Fry1989, Herbythyme, Homo lupus, Klemen Kocjancic, Krun, Liftarn, Mattes, Neq00, Nightstallion, Pumbaa80, Trockennasenaffe, 5 anonymous edits

File:Flag of Tuvalu.svg *Source*: http://en.wikipedia.org/w/index.php?title=File:Flag_of_Tuvalu.svg *License*: unknown *Contributors*: Alkari, Avala, CommonsDelinker, Davepape, Denelson83, DiamondVertex, Fry1989, Greentubing, Homo lupus, Jed, Klemen Kocjancic, Kookaburra, Mattes, Neq00, Nightstallion, Vzb83, 8 anonymous edits

File:Flag of Wallis and Futuna.svg *Source*: http://en.wikipedia.org/w/index.php?title=File:Flag_of_Wallis_and_Futuna.svg *License*: unknown *Contributors*: User:Denelson83

File:Map of Oceania-Pacific wide.svg *Source*: http://en.wikipedia.org/w/index.php?title=File:Map_of_Oceania-Pacific_wide.svg *License*: unknown *Contributors*: User:AkBowl, User:Sémhur, User:Tintazul

Image:Fiji Cook Island rugby.jpg *Source*: http://en.wikipedia.org/w/index.php?title=File:Fiji_Cook_Island_rugby.jpg *License*: unknown *Contributors*: learza

Image:Australian World Cup treble.jpg *Source*: http://en.wikipedia.org/w/index.php?title=File:Australian_World_Cup_treble.jpg *License*: unknown *Contributors*: Matt Sterne

File:View of part of the town of Pelew cph.3b32619.jpg *Source*: http://en.wikipedia.org/w/index.php?title=File:View_of_part_of_the_town_of_Pelew_cph.3b32619.jpg *License*: unknown *Contributors*: Engraving by W. & J. Walker after drawing by T. Maltron after sketch by Robert White.

File:Palau-Inseln.jpg *Source*: http://en.wikipedia.org/w/index.php?title=File:Palau-Inseln.jpg *License*: unknown *Contributors*: LSDSL, Telim tor

Image:Koror in the Japanese Period.JPG *Source*: http://en.wikipedia.org/w/index.php?title=File:Koror_in_the_Japanese_Period.JPG *License*: unknown *Contributors*: User:

Image:Capitol-complex-melekeok-palau20071220 crop.jpg *Source*: http://en.wikipedia.org/w/index.php?title=File:Capitol-complex-melekeok-palau20071220_crop.jpg *License*: unknown *Contributors*: Photographer Binter; Original uploader was Binter at de.wikipedia (Original text : Peter Binter)

MIX
Papier aus verantwortungsvollen Quellen
Paper from responsible sources
FSC® C105338

Printed by Books on Demand GmbH, Norderstedt / Germany